ISSN 0044-2267

ZEITSCHRIFT FÜR ANGEWANDTE MATHEMATIK UND MECHANIK

INGENIEURWISSENSCHAFTLICHE FORSCHUNGSARBEITEN

BAND 61 1981 HEFT 2

AUS DEM INHALT

HAUPTAUFSÄTZE

THEORIA CUM PRAXI

AKADEMIE-VERLAG · BERLIN

ZAMM Bd. 61 Nr. 2 S. 69—132 Berlin, Februar 1981

EVP 18,— M 34115

INHALT

Wir bitten, **Manuskriptsendungen** zweifach (Original und eine Kopie, sprachlich einwandfrei, Formeln mit Maschine oder in Druckschrift geschrieben) an folgende Anschrift zu richten:

Zeitschrift für Angewandte Mathematik und Mechanik, z. Hd. Herrn Prof. Dr. Günter Schmidt
Institut für Mechanik der Akademie der Wissenschaften der DDR
DDR-1199 Berlin, Rudower Chaussee 5

Zu den Arbeiten, die als Hauptaufsätze bestimmt sind, ist auf gesondertem Blatt eine Zusammenfassung von 5 bis 10 Zeilen in englischer und (möglichst) deutscher und russischer Sprache beizufügen. Ausführliche Hinweise für die Autoren, um deren strikte Berücksichtigung gebeten wird, finden sich im Anschluß an das Inhaltsverzeichnis des Jahrganges **59** (1979).

Die Autoren erhalten von den Hauptaufsätzen 75, von den Kleinen Mitteilungen 25 Sonderdrucke ohne Berechnung, darüber hinaus weitere Sonderdrucke gegen Berechnung.

Der Verlag behält sich für alle Beiträge das Recht der Vervielfältigung, Vorbereitung und Übersetzung vor.

Bestellungen sind zu richten

- **in der DDR** an den Postzeitungsvertrieb unter Angabe der Kundennummer des Bestellenden oder an den AKADEMIE-VERLAG, DDR-1080 Berlin, Leipziger Str. 3—4
- **im sozialistischen Ausland** an eine Buchhandlung für fremdsprachige Literatur oder an den zuständigen Postzeitungsvertrieb
- **in der BRD und Berlin(West)** an eine Buchhandlung oder an die Auslieferungsstelle KUNST UND WISSEN, Erich Bieber OHG, D-7000 Stuttgart 1, Wilhelmstraße 4—6
- **in Österreich** an den Globus-Buchvertrieb, A-1201 Wien, Höchstädtplatz 3
- **in den übrigen westeuropäischen Ländern** an eine Buchhandlung oder an die Auslieferungsstelle KUNST UND WISSEN, Erich Bieber GmbH, Ch-8008 Zürich/Schweiz, Dufourstraße 51
- **im übrigen Ausland** an den Internationalen Buch- und Zeitschriftenhandel; den Buchexport, Volkseigener Außenhandelsbetrieb der Deutschen Demokratischen Republik, DDR-7010 Leipzig, Postfach 160; oder an den AKADEMIE-VERLAG, DDR-1080 Berlin, Leipziger Straße 3—4

ZEITSCHRIFT FÜR ANGEWANDTE MATHEMATIK UND MECHANIK

Herausgeber und Chefredakteur: Prof. Dr. Günter Schmidt. Redaktion: Dr. Winfried Heinrich, Dr. Horst Weinert, Dipl.-Math. Friedhild Dudel, Helga Rühl, Institut für Mechanik der Akademie der Wissenschaften der DDR. Verlag: Akademie-Verlag, DDR-1080 Berlin, Leipziger Straße 3—4; Fernruf: 2236221 oder 2236229. Telex-Nr.: 114420; Bank: Staatsbank der DDR, Berlink Kto.-Nr.: 6836-26-20712. Aschrift der Redaktion: Institut für Mechanik der Akademie der Wissenschaften, DDR-1199 Berlin, Rudower Chaussee 5, Fernruf: 6802841/3639.

Veröffentlicht unter der Lizenznummer 1282 des Presseamtes beim Vorsitzenden des Ministerrates der Deutschen Demokratischen Republik. Gesamtherstellung: VEB Druckerei „Thomas Müntzer", 5820 Bad Langensalza. Erscheinungsweise: Die Zeitschrift für Angewandte Mathematik und Mechanik erscheint monatlich. Die 12 Hefte eines Jahres einschließlich Tagungsheft bilden einen Band. Bezugspreis je Band 360,— M zuzüglich Versandspesen (Preis für die DDR 216,— M). Bezugspreis je Heft 30,— M (Preis für die DDR 18,— M) Bestellnummer dieses Heftes: 1009/61/2.

AN (EDV) 35937

ZEITSCHRIFT FÜR ANGEWANDTE MATHEMATIK UND MECHANIK

INGENIEURWISSENSCHAFTLICHE FORSCHUNGSARBEITEN

HERAUSGEGEBEN VON G. SCHMIDT, BERLIN

Band 60

sechzigster Jahrgang 1980

AKADEMIE-VERLAG · BERLIN

Inhaltsübersicht

A. Verfasserverzeichnis

(B = Bericht; Ber. = Berichtigung; H = Hauptaufsatz; HV = Hauptvortrag; KM = Kleine Mitteilung; N = Nachricht; V = Vortragsauszug)

Der Buchstabe T vor der Seitenzahl weist darauf hin, daß der betreffende Beitrag in den Sonderheften 6 und 7 (GAMM-Tagung 1980) erschienen ist.

ZEITSCHRIFT FÜR ANGEWANDTE MATHEMATIK UND MECHANIK

Herausgeber und Chefredakteur: Prof. Dr. Günter Schmidt (Redakteure: Dr. W. Heinrich und Dr. H. Weinert), Institut für Mechanik der Akademie der Wissenschaften der DDR. Verlag: Akademie-Verlag, DDR-1080 Berlin, Leipziger Straße 3–4; Fernruf: 2236229 oder 2236221; Telex-Nr.: 114420; Bank: Staatsbank der DDR, Berlin, Kto.-Nr.: 6836-26-20712. Anschrift der Redaktion: Institut für Mechanik der Akademie der Wissenschaften der DDR, DDR-1199 Berlin, Rudower Chaussee 5; Fernruf 6702841/3639 u. 3643.

Veröffentlicht unter der Lizenznummer 1282 des Presseamtes beim Vorsitzenden des Ministerrates der Deutschen Demokratischen Republik. Gesamtherstellung: VEB Druckerei „Thomas Müntzer", DDR-5820 Bad Langensalza. Erscheinungsweise: Die Zeitschrift für Angewandte Mathematik und Mechanik erscheint monatlich. Die 12 Hefte eines Jahres einschließlich Tagungsheft bilden einen Band. Bezugspreis je Band 324,– M zuzüglich Versandspesen (Preis für die DDR 216,– M). Bezugspreis je Heft 27,– M (Preis für die DDR 18,– M).

AN (EDV) 35937

B. Sachverzeichnis

(Abkürzungen siehe Seite II)

Seite

Approximationstheorie

(siehe auch Numerische Mathematik und Rechentechnik)

Differential- und Integralgleichungen

(siehe auch Numerische Mathematik und Rechentechnik; Operationsforschung, Optimierung, optimale Prozesse; Regelungssysteme, Automatentheorie)

Dynamik, Schwingungen

(siehe auch Differential- und Integralgleichungen; Elasto- und Plastomechanik; Mechanik (allgemein) und Physik; Operationsforschung, Optimierung, optimale Prozesse; Strukturmechanik (Stäbe, Platten, Schalen usw.))

Elasto- und Plastomechanik

(siehe auch Strukturmechanik (Stäbe, Platten, Schalen usw.); Wärmeübertragung)

Gasdynamik und Magnetohydrodynamik

Seite

Grenzschichttheorie und Turbulenz

Intervallmathematik

Mathematik, allgemein

Seite

Matrizen

Mechanik (allgemein) und Physik

(siehe auch Differential- und Integralgleichungen; Dynamik, Schwingungen; Elasto- und Plastomechanik; Gasdynamik und Magnetohydrodynamik; Grenzschichttheorie und Turbulenz; Strukturmechanik (Stäbe, Platten, Schalen usw.); Strömungsmechanik

Numerische Mathematik und Rechentechnik

(siehe auch Approximationstheorie; Differential- und Integralgleichungen; Grenzschichttheorie und Turbulenz; Intervallmathematik; Mechanik (allgemein) und Physik; Operationsforschung, Optimierung, optimale Prozesse; Strömungsmechanik; Wahrscheinlichkeitstheorie und Mathematische Statistik)

Operationsforschung, Optimierung, optimale Prozesse

Regelungssysteme, Automatentheorie

(siehe auch Dynamik, Schwingungen)

Seite

Strukturmechanik (Stäbe, Platten, Schalen usw.)

(siehe auch Dynamik, Schwingungen; Elasto- und Plastomechanik)

Strömungsmechanik

(siehe auch Dynamik, Schwingungen; Gasdynamik und magnetohydrodynamik; Grenzschichttheorie und Turbulenz; Numerische Mathematik und Rechentechnik)

Wärmeübertragung

(siehe auch Elasto- und Plastomechanik; Gasdynamik und Magnetohydrodynamik; Numerische Mathematik und Rechentechnik)

Wahrscheinlichkeitstheorie und Mathematische Statistik

(siehe auch Numerische Mathematik und Rechentechnik)

Wissenschaftliche Gesellschaften, Tagungen, Nachrichten

C. Buchbesprechungen

Hinweise für Autoren / Remarks for authors

Manuskripte (zweifach, Original und eine Kopie) und andere den Inhalt der Zeitschrift betreffende Zuschriften sind dem Herausgeber der ZAMM entweder über ein Mitglied des Herausgebergremiums oder direkt über folgende Anschrift zuzuleiten:

Zeitschrift für Angewandte Mathematik und Mechanik
Institut für Mechanik
der Akademie der Wissenschaften der DDR
DDR-1199 Berlin, Rudower Chaussee 5

Bei Einreichen des Manuskripts hat der Autor verbindlich zu erklären, daß sein Aufsatz noch in keiner anderen allgemein zugänglichen Zeitschrift veröffentlicht worden ist bzw. kein Antrag auf Veröffentlichung in einer solchen Zeitschrift läuft.
Es wird erwartet, daß der Autor über eine weitere Kopie verfügt.
Die Manuskripte müssen in ihrer äußeren Gestaltung folgenden Anforderungen genügen, anderenfalls sie zurückgewiesen werden können oder in der Bearbeitung lange Verzögerungen eintreten:
1. Die Manuskripte sind in einwandfreier sprachlicher Form (einschließlich Orthographie, Interpunktion) einzureichen.
2. Die Manuskriptseiten sind nur einseitig zu beschreiben. Der Text ist mit Maschine mit weitem Zeilenabstand (zweizeilig) und mit mindestens 3 cm breitem Rand zu schreiben.
3. Die **Formeln** müssen einwandfrei lesbar mit Maschine oder in Druckschrift und mit großem Zeilenabstand geschrieben sein. Für mehrfach wiederkehrende umfangreiche oder drucktechnisch komplizierte Terme sind abkürzende Bezeichnungen einzuführen. Bei Exponentialausdrücken mit längerem Exponenten ist weitgehend von der Schreibweise exp (...) Gebrauch zu machen. Die verwendeten Formelzeichen sind in einer für den Setzer eindeutigen Weise zu kennzeichnen und gegebenenfalls in einer besonderen Liste zu erläutern. Um unnötige Korrekturen zu vermeiden, ist genau zu unterscheiden, zwischen v, ν, γ und r; φ und Φ; ψ und Ψ; 0(Null, O, o, Θ und θ $\times$; 1, l und e; K, k und $\varkappa$; i und ι; ζ und ξ; x, X und χ; u und n. Griechische Buchstaben können rot unterstrichen werden. Die wichtigsten, insbesondere auch die im weiteren Verlauf des Textes zitierten Formeln sind am **rechten** Rand der Manuskriptseiten in runden Klammern durchzunumerieren.

4. Es ist darauf zu achten, daß alle benutzten Abkürzungen definiert sind. Es wird gebeten, Textabkürzungen wie Dgl., eq. zu vermeiden.
Sätze, Hilfssätze, Korrolare, Definitionen, Beweise u. dgl. sind dadurch hervorzuheben, daß ihnen in Sperrdruck das betreffende Wort (Satz . . ., Beweis . . ., usw.) vorangestellt wird. Die Aussage eines Satzes oder Hilfssatzes wird einheitlich kursiv gedruckt. Das Ende einer solchen Aussage oder eines Beweises ist im Manuskript zu markieren. Bei Hauptaufsätzen ist im Interesse einer übersichtlicheren Gliederung und besseren Lesbarkeit weitgehend von der Möglichkeit Gebrauch zu machen, einzelne Passagen (Beweise, Bemerkungen, . . .(durch Kleindruck gegenüber dem Haupttext zurücktreten zu lassen und die Stellen durch den Hinweis „Petit" zu kennzeichnen.
5. Als Vorlagen für die **Abbildungen** sind saubere Zeichnungen in Bleistift oder Tusche (Beschriftung nur mit Bleistift) auf gesonderten Blättern und in einem genügend großen, eine Verkleinerung gestattenden Format erforderlich. Da Korrekturen in den Abbildungen schwierig sind, ist hier besonders sorgfältig zwischen ähnlichen Buchstaben zu unterscheiden (vgl. 3).
6. Auf **Literatur** ist im Text durch Nummern in eckigen Klammern zu verweisen. Das nach diesen Nummern geordnete Literaturverzeichnis erscheint am Ende des Aufsatzes. Es wird gebeten, darin die Literatur nach folgendem Muster anzuführen:
bei Büchern:
HOUSEHOLDER, A. S., The theory of matrices in numerical analysis, Blaisdell Publ. Comp., New York/Toronto/London 1964.
bei Zeitschriftenartikeln:
WEISSINGER, J., Zur nichtlinearen Theorie der ungleichförmigen Umströmung von Profilen, Z. Angew. Math. u. Mech. **50** (1970), 337–346.
bei Sammelwerken:
BARBER, J. R., Thermoelastic contact problems. In: DE PATER, A. D.; KALKER, J. J. (Ed.), The mechanics of the contact between deformable bodies. Proc. Symp. IUTAM, Enschede 20. bis 23. 8. 74. Delft: Delft University Press **1975**, S. 177–205.
Russischsprachige Zitate können in kyrillischen Buchstaben oder in einer der gebräuchlichen Transliterationen geschrieben werden.
Handelt es sich bei der zitierten Literaturstelle um eine Übersetzung, so ist anzugeben, wo bzw. in welcher Zeitschrift, in welchem Jahr und in welcher Sprache das Original erschienen ist.
7. Den Manuskripten, die als Hauptaufsätze bestimmt sind, ist auf gesondertem Blatt eine **Zusammenfassung** von höchstens 15 Zeilen in englischer, deutscher und russischer Sprache beizufügen. Falls sie nicht in deutscher oder russischer Sprache geliefert werden kann, ist wenigstens eine Liste der vorkommenden Fachausdrücke in der fehlenden Sprache erwünscht.
Auch bei Kleinen Mitteilungen ist eine kurze Zusammenfassung in der Sprache der Arbeit erwünscht.
8. Dem Manuskript ist am Ende die **Anschrift der Autoren** mit akademischem Titel und Institut) beizufügen, die am Ende des Aufsatzes zur Information des Lesers veröffentlicht wird. Darüber hinaus sind diejenigen Anschriften anzugeben, an welche die Korrekturabzüge und die Sonderdrucke zu senden sind. Ändert sich während der Laufzeit einer Arbeit eine Anschrift, so ist dies zur Vermeidung von Fehlleitungen und langen Verzögerungen dem Herausgeber der ZAMM mitzuteilen. Die Autoren erhalten von den Hauptaufsätzen 75, von den Kleinen Mitteilungen 25 Sonderdrucke ohne Berechnung darüber hinaus weitere gegen Berechnung.

Manuscripts (master and one copy) and other correspondence relating to the contents of the journal should be addressed to the editor of ZAMM, either through any one member of the editorial board or direct to the following address:

Zeitschrift für Angewandte Mathematik und Mechanik
Institut für Mechanik
der Akademie der Wissenschaften der DDR
DDR-1199 Berlin, Rudower Chaussee 5

On submitting manuscripts authors must give the explicit assurance that the paper submitted has not previously been published in, nor been submitted for publication to, another public journal.
Authors should have a copy of their manuscript.
The following rules in the set-up of manuscripts should be observed to avoid the risk of delayed publication or even rejection.
1. The manuscripts should be well written (including spelling and punctuation).
2. The text should be typewritten using only one side of the sheet, with adequate line spacing (2-line spacing) and a margin of not less than 3 cm.
3. **Formulas** must be clearly legible (typewritten or written clearly) and with adequate line spacing. Long terms that occur especially frequently or terms that present especial difficulty to type-setters should be abbreviated, if possible. In exponential expressions with comparatively long exponents the abbreviation "exp(. . .)" should preferably be used throughout. Signs used in formulas should be clearly recognizable to type-setters or, if necessary, explained in an appended list. To avoid tiresome corrections special care should be taken to clearly distinguish between: v, γ, v and r; φ and Φ; ψ and Ψ; 0(zero), O, o, Θ and Θ; l, l und e, K, k and χ; i and i; ζ and ξ; x, X and χ; u and n. It would be helpful if Greek letters are underlined in red. Major formulas, in particular the ones quoted throughout the text, should be numbered continually on the right margin (in parentheses).
4. All abbreviations must be defined, and abbreviated words or text sections should be avoided.
Theorems, corollaries, definitions, proofs, etc. should be emphasized by printing, in spaced type, the relevant term (theorem . ., proof . . ., etc.) in front of the expression, which, in its turn, will be printed uniformly in italics. The end of a theorem, proof. etc. should be marked in the manuscript. Certain passages (proofs, remarks, etc.) in articles and monographs may be printed in smaller type to ensure easy readability and should, therefore, be marked "petit" in the manuscript.
5. **Illustations** should be in the form of clear pencil or ink drawings (captions, legends, etc. in pencil only!) on separate sheets, big enough to be scaled down.
Since corrections in illustrations are difficult special care must be taken to differentiate clearly between similar letters (see paragraph 3).
6. **References** should be numbered in brackets throughout the text, with the bibliography according to these numbers appended. References should follow the model printed below:
books:
HOUSEHOLDER, A. S., The theory of matrices in numerical analysis, Blaisdell Publ. Comp., New York/Toronto/London 1964, p. 74.
articles:
WEISSINGER, J., Zur nichtlinearen Theorie der ungleichförmigen Umströmung von Profilen, Z. Angew. Math. u. Mech. **50** (1970), 337–346.
proceedings etc.:
BARBER, J. R., Thermoelastic contact problems. In: DE PATER, A. D.; KALKER, J. J. (Ed.), The mechanics of the contact between deformable bodies. Proc. Symp. IUTAM, Enschede 20.–23. 8. 74. Delft: Delft University Press 1975, S. 177–205.
Russian quotations can be written in Cyrillic letters or in customary transcriptions.
In cases of translated references, the original source (name of periodical, year of publication, original language) should be stated.
7. A concise **summary** (not more than 15 lines) in English, German or Russian should be provided on a separate sheet for manuscripts intended as major articles. If a summary cannot be provided in German or Russian, a list of the technical terms of the missing language will be appreciated. A brief summary is also desired of shorter contributions (Kleine Mitteilungen) in the same language as the contribution.
8. Authors should state their academic title, institute and address which will be published at the end of the contribution for the information or readers. Addresses should also be stated of persons to whom galley proofs and reprints are to be sent. The editors of ZAMM should be notified of any change of the author's address occuring after submission of a contribution, so as to avoid any postal errors and delays. Authors of major articles will be supplied with 75 free reprints: authors of other contributions will receive 25. All additional reprints have to be paid for.

ZAMM 61, 69 – 74 (1981)

P. RUGE

Ein Variationskonzept zur Reduktion statischer Stabilitätsprobleme

Meinem verehrten Lehrer, Herrn Prof. Dr.-Ing. habil. S. FALK, zum 60. Geburtstag am 6. Mai 1981 in Dankbarkeit gewidmet.

Im Rahmen der Näherungsberechnung von statischen Stabilitätslasten λ für kontinuierliche Systeme wird gezeigt, wie die Ordnung der finiten Übersetzung $\boldsymbol{Ax} = \lambda \boldsymbol{Bx}$ reduziert werden kann. Die Methode beruht auf der Einführung von Versatzmomenten infolge der Spannungen im Grundzustand als unabhängiger Zustandsgröße neben der Durchsenkung. Dieses Variationskonzept erlaubt eine exakte statische Kondensation durch Elimination der Weggrößenparameter.

A method for reducing the size of finite element systems $\boldsymbol{Ax} = \lambda \boldsymbol{Bx}$ in structural stability is presented. The technique is based upon a variational concept in which the displacements and distributed moments caused by the in-plane loads λ are taken to be independent. This theorem allows an accurate static condensation by an elimination of the displacement parameters.

В рамках приближённого расчёта статистических нагрузок устойчивости λ для непрерывных систем показывается, как можно понизить порядок системы конечных элементов $\boldsymbol{Ax} = \lambda \boldsymbol{Bx}$.
Метод основан на введении моментов сдвига в результате напряжений в основном состоянии как независимой величины состояния наряду с прогибом. Этот вариационный метод позволяет точную статическую конденсацию путём исключения параметров перемещения.

1. Einführung

Zur näherungsweisen Lösung der Zustandsgleichungen linear elastischer Kontinua über finite Übersetzungen steht eine Vielzahl von zumindest stationären Funktionalen zur Verfügung, die eine synchrone Approximation mehrerer Zustandsgrößen bei weitgehender Freiheit in den Ansätzen erlauben. Eine umfassende Systematik sogenannter dual-komplementärer Variationsprinzipien der mathematischen Physik findet man bei ODEN und REDDY [1]. Besonders für die statische Berechnung auch verwickelter Strukturen liegen vielfältige numerische Erfahrungen vor.

Die Berechnung von Eigenfrequenzen ω und Knicklasten zum Beispiel eines BERNOULLIbalkens erfordert neben einer zweckmäßigen Formulierung der Formänderungsenergie eine Darstellung der kinetischen Energie E_{kin} beziehungsweise der Arbeit A_{ax} der Axialkräfte H, die überwiegend als Weggrößenfunktionale ausgewertet werden:

$$E_{\text{kin}} = \tfrac{1}{2} \int \varrho F \dot{w}^2(x, t)\, \mathrm{d}x \to -\tfrac{1}{2} \omega^2 \int \varrho F w^2(x)\, \mathrm{d}x\,, \qquad w(x, t) = w(x)\, \mathrm{e}^{i\omega t}\,, \tag{1}$$

$$A_{ax} = -\tfrac{1}{2} \int H w_{,x}^2\, \mathrm{d}x; \qquad (\ )_{,x} = \mathrm{d}(\ )/\mathrm{d}x\,, \qquad (\ )^{\cdot} = \mathrm{d}(\ )/\mathrm{d}t\,. \tag{2}$$

F steht dabei für die Querschnittsfläche, ϱ für die spezifische Massendichte, w bezeichnet die Balkendurchsenkung, x die Balkenkoordinate, t die Zeit.

Eine komplementäre kinetische Energie erhält man durch Einführung eines „Winkelimpulses" ϱp nach TABARROK [2], dessen zeitliche Ableitung das Moment M darstellt:

$$M(x, t) = (\varrho p)^{\cdot}\,. \tag{3}$$

Der Schwerpunktsatz für ein Balkenelement liefert mit (3) eine Momenten-Geschwindigkeitsbeziehung ($v = \dot{w}$),

$$M_{,xx} = \varrho F \ddot{w} = (\varrho F v)^{\cdot} = \varrho \dot{p}_{,xx} \to Fv = p_{,xx}, \tag{4}$$

die zur Elimination der Geschwindigkeit v benutzt werden kann,

$$E_{\text{kin}} = \tfrac{1}{2} \int \varrho F^{-1} p_{,xx}^2\, \mathrm{d}x\,, \tag{5}$$

und zusammen mit der Formänderungsenergie

$$E_{FÄ} = \frac{1}{2} \int \frac{M^2}{EJ}\, \mathrm{d}x = \frac{1}{2} \int \frac{(\varrho \dot{p})^2}{EJ}\, \mathrm{d}x \to -\frac{\omega^2}{2} \int \frac{\varrho^2 p^2}{EJ}\, \mathrm{d}x\,, \qquad p(x, t) = p(x)\, \mathrm{e}^{i\omega t}\,, \tag{6}$$

eine Diskretisierungsgrundlage darstellt. Numerische Beispiele findet man bei TABARROK selbst und auch bei GERADIN [3]. Die Einführung der Geschwindigkeit $v = \dot{w}$ als eigener Zustandsgröße und eine Aufspaltung der Energie in zwei Teile

$$E_{\text{kin}} = -\tfrac{1}{2} \int \varrho F v^2\, \mathrm{d}x + i\omega \int \varrho F w v\, \mathrm{d}x\,, \qquad i = \sqrt{-1}\,, \tag{7}$$

ermöglicht eine unabhängige Variation der Geschwindigkeit ohne Berücksichtigung irgendwelcher Rand- und Übergangsbedingungen. Die Verknüpfung der Eigenfrequenz ω allein mit dem bilinearen Term in (7) erlaubt eine wahlweise Elimination der Weggrößen- oder Geschwindigkeitsparameter, was zu einer erheblichen Reduzierung der Ordnung der finiten Übersetzung als Matrizen-Eigenwertaufgabe $(\boldsymbol{A} - \omega^2 \boldsymbol{B})\, \boldsymbol{x} = \boldsymbol{0}$ führen kann. Das skizzierte Konzept wurde von HUGHES [4, 5] ausführlich behandelt und numerisch erprobt.

2. Arbeit der Axialkraft

Eine ähnliche Aufspaltung wie für die kinetische Energie in (7) bietet sich nun für die Arbeit A_{ax} (2) der Axialkräfte $H(x)$ an, wobei das Versatzmoment

$$m = Hw_{,x} \tag{8}$$

der Horizontalkraft selbst oder die modifizierte Form

$$m^* = \sqrt{H}\, w_{,x} \tag{9}$$

geeignete selbständige Zustandsgrößen sind und folgende gemischte Funktionale definieren:

$$G^* = \tfrac{1}{2} \int EJw_{,xx}^2 \,\mathrm{d}x - \int \sqrt{H}\, w_{,x}m \,\mathrm{d}x + \tfrac{1}{2} \int m^2 \,\mathrm{d}x\,, \tag{10}$$

$$G = \tfrac{1}{2} \int EJw_{,xx}^2 \,\mathrm{d}x - \int w_{,x}m \,\mathrm{d}x + \tfrac{1}{2} \int H^{-1}m^2 \,\mathrm{d}x\,, \qquad H \neq 0\,. \tag{11}$$

In der zweiten Fassung (11) darf die Axialkraft $H(x)$ in keinem Punkt des Integrationsgebietes Null werden.

Das Verschwinden der Variation $\delta_m G$ des Funktionals G (11) bezüglich der Versatzmomente m,

$$\delta_m G = \int (w_{,x} - H^{-1}m)\, \delta m \,\mathrm{d}x \overset{!}{=} 0 \rightarrow m = w_{,x}H\,, \tag{12}$$

als notwendiger Bedingung für den zumindest stationären Charakter des Funktionals führt auf die Interpretation (8) der Zustandsgröße m bei beliebiger Variation δm.

Das Verschwinden der Variation $\delta_w G$ bezüglich der Verschiebungen w,

$$\delta_w G = [EJw_{,xx}\, \delta w_{,x}] - [(EJw_{,xx})_{,x} - m]\, \delta w + \int [(EJw_{,xx})_{,xx} + m_{,x}]\, \delta w \,\mathrm{d}x \overset{!}{=} 0\,, \tag{13}$$

führt nach zweimaliger partieller Integration mit (12) auf die Differentialgleichung der Balkenknickung samt einiger Randterme als Produkte der konjugierten Größen aus Durchsenkung w und Querkraft Q sowie aus Neigung $w_{,x}$ und Moment M:

$$M = EJw_{,xx}\,, \qquad Q = (EJw_{,xx})_{,x} - Hw_{,x}\,. \tag{14}$$

Die Randterme in (13) und damit die Variation $\delta_w G$ sind identisch Null, falls die Ansatzfunktionen die geometrischen Rand- und Übergangsbedingungen erfüllen, also sogenannte zulässige Näherungen sind.

3. Finite Übersetzung und Reduktion

Die finite Übersetzung mit anschließender Reduktion wird am Beispiel des Funktionals G (11) vorgeführt.

Zunächst werden einige dimensionslose Parameter und Funktionen mit der Bereichslänge l und konstanten Bezugsgrößen EJ. und H. für die Biegesteifigkeit und die Axialkraft vereinbart,

$$\begin{gathered} x = l\xi\,, \qquad 0 \leqq \xi \leqq 1\,, \qquad \mathrm{d}(\,)/\mathrm{d}\xi = (\,)_{,1}\,, \qquad \hat{w} = w/l\,, \\ H(\xi) = H_0 h(\xi)\,, \qquad EJ(\xi) = EJ_0 g(\xi)\,, \qquad \hat{m} = \frac{ml^2}{EJ_0}\,, \qquad \lambda = \frac{H_0 l^2}{EJ_0}\,, \end{gathered} \tag{15}$$

die das Funktional (11) in eine für die Auswertung geeignete Form überführen:

$$\hat{G} = \frac{Gl}{EJ_0} = \frac{1}{2} \int g\hat{w}_{,11}^2 \,\mathrm{d}\xi - \int \hat{w}_{,1}\hat{m} \,\mathrm{d}\xi + \frac{1}{2} \int \frac{\hat{m}^2}{h\lambda} \,\mathrm{d}\xi\,. \tag{16}$$

Getrennte Ansätze

$$\hat{w}(\xi) = \boldsymbol{w}^T \boldsymbol{\varphi}(\xi)\,, \qquad \hat{m}(\xi) = \boldsymbol{m}^T \boldsymbol{\psi}(\xi) \tag{17}$$

für die Verschiebungen und die Versatzmomente, zweckmäßig mit normierten Koordinatenfunktionen $\boldsymbol{\varphi}$ und $\boldsymbol{\psi}$ in Form von HERMITEpolynomen und den dazugehörigen Parameterspalten $\boldsymbol{w}$ und $\boldsymbol{m}$, übersetzen das Funktional G (11) in eine Summe quadratischer beziehungsweise bilinearer Formen

$$\hat{G} = \tfrac{1}{2} \boldsymbol{w}^T \boldsymbol{S}_w \boldsymbol{w} - \boldsymbol{w}^T \boldsymbol{G} \boldsymbol{m} + \tfrac{1}{2} \Lambda^{-1} \boldsymbol{m}^T \boldsymbol{S}_m \boldsymbol{m} \tag{18}$$

mit dem Näherungs-Beulwert Λ sowie symmetrischen Matrizen $\boldsymbol{S}_w$, $\boldsymbol{S}_m$ und einer in der Regel nicht quadratischen Matrix $\boldsymbol{G}$:

$$\boldsymbol{S}_w = \int g \boldsymbol{\varphi}_{,11} \boldsymbol{\varphi}_{,11}^T \,\mathrm{d}\xi\,, \qquad \boldsymbol{S}_m = \int h^{-1} \boldsymbol{\psi}\boldsymbol{\psi}^T \,\mathrm{d}\xi\,, \qquad \boldsymbol{G} = \int \boldsymbol{\varphi}_{,1} \boldsymbol{\psi}^T \,\mathrm{d}\xi\,. \tag{19}$$

Im Extremalpunkt des angenäherten Funktionals (18) verschwindet der Gradient bezüglich der Parameter $\boldsymbol{m}$ und $\boldsymbol{w}$, wodurch eine homogene Matrizenaufgabe zur Bestimmung der Näherungsknickwerte Λ für λ entsteht:

$$\underset{\begin{bmatrix} \boldsymbol{w} \\ \boldsymbol{m} \end{bmatrix}}{\operatorname{grad}} \hat{G} = \begin{bmatrix} \boldsymbol{S}_w & -\boldsymbol{G} \\ -\boldsymbol{G}^T & \Lambda^{-1}\boldsymbol{S}_m \end{bmatrix} \begin{bmatrix} \boldsymbol{w} \\ \boldsymbol{m} \end{bmatrix} = \boldsymbol{0}\,. \tag{20}$$

Eine der Blockzeilen in (20) kann zur Elimination einer der Parameterspalten herangezogen werden. In der Regel ist die Approximation der Biegelinie anspruchsvoller, da die geometrischen Randbedingungen zu erfüllen sind und die Formänderungsenergie die Ansatzfunktionen $\boldsymbol{\varphi}$ nur in stark aufgerauhter Form als zweite Ableitungen verwertet. Als Folge davon wird die Anzahl n_w der Ansatzparameter $\boldsymbol{w}$ die Zahl n_m der Parameter $\boldsymbol{m}$ deutlich übertreffen, so daß sich die Elimination der Weggrößen anbietet.

$$(20): \quad \boldsymbol{w} = \boldsymbol{S}_w^{-1}\boldsymbol{G}\boldsymbol{m} \,. \tag{21}$$

Das verbleibende Eigenwertproblem folgt durch Einsetzen von (21) in die untere Blockzeile von (20):

$$[\boldsymbol{S}_m - \Lambda \boldsymbol{G}^T\boldsymbol{S}_w^{-1}\boldsymbol{G}]\,\boldsymbol{m} = \boldsymbol{0}\,, \qquad \Lambda = \frac{H_0 l^2}{EJ_0}. \tag{22}$$

Zur numerischen Auswertung des Matrizenproduktes in (22) empfiehlt sich eine Choleskyzerlegung für $\boldsymbol{S}_w$, so daß ein Produkt aus zwei gleichen Matrizen zu berechnen bleibt:

$$\boldsymbol{S}_w = \boldsymbol{R}\boldsymbol{R}^T \rightarrow \boldsymbol{G}^T\boldsymbol{S}_w^{-1}\boldsymbol{G} = (\boldsymbol{R}^{-1}\boldsymbol{G})^T\,(\boldsymbol{R}^{-1}\boldsymbol{G})\,. \tag{23}$$

Der zugeordnete Rayleighquotient

$$R = \frac{\boldsymbol{z}^T\boldsymbol{S}_m\boldsymbol{z}}{\boldsymbol{z}^T\boldsymbol{G}^T\boldsymbol{S}_w^{-1}\boldsymbol{G}\boldsymbol{z}}\,, \qquad R_{\text{Extr.}} = \Lambda_i\,, \qquad \boldsymbol{z}_{\text{Extr.}} = \boldsymbol{m}_i\,, \tag{24}$$

zum Eigenwertproblem (22) ist ein Minimalquotient bezüglich des kleinsten Beulwertes, da die inverse Formänderungsenergie des Balkenkontinuums normiert mit dem Skalar $\boldsymbol{z}^T\boldsymbol{S}_m\boldsymbol{z}$ im Nenner erscheint, so daß der exakte Knickwert λ wie beim reinen Weggrößenfunktional von oben angenähert wird,

$$\lambda_1 \leqq \Lambda_1\,. \tag{25}$$

4. Bezug zum reinen Weggrößenfunktional

Bei bekannten Eigenvektoren $\boldsymbol{m}$ der reduzierten Aufgabe (22) folgen die zugeordneten Weggrößenparameter $\boldsymbol{w}$ aus der Eliminationsgleichung (21). Es liegt nahe, diese Vektoren $\boldsymbol{w}$ als Näherungen für die Matrizeneigenwertaufgabe

$$(\boldsymbol{S}_w - \tilde{\Lambda}\boldsymbol{H})\,\tilde{\boldsymbol{w}} = \boldsymbol{0}\,, \qquad \boldsymbol{H} = \int h\boldsymbol{\varphi}_{,1}\boldsymbol{\varphi}_{,1}^T\,\mathrm{d}\xi\,, \tag{26}$$

zu benutzen, die aus der finiten Übersetzung des reinen Weggrößenfunktionals mit der Arbeit A_{ax} (2) der Axialkräfte folgt. Zur Unterscheidung von den Eigenlösungen Λ, $\boldsymbol{m}$, $\boldsymbol{w}$ der reduzierten Aufgabe (22) sind die entsprechenden Größen in (26) mit einer Tilde versehen.

Der zugeordnete Rayleighquotient

$$R_i = \frac{Z_R(\boldsymbol{w}_i)}{N_R(\boldsymbol{w}_i)} = \frac{\boldsymbol{w}_i^T\boldsymbol{S}_w\boldsymbol{w}_i}{\boldsymbol{w}_i^T\boldsymbol{H}\boldsymbol{w}_i}\,, \qquad \boldsymbol{w} = \boldsymbol{S}_w^{-1}\boldsymbol{G}\boldsymbol{m}\,, \tag{27}$$

zur Aufgabe (26) nähert den exakten Beulwert λ weiterhin von oben an,

$$\lambda_1 \leqq \tilde{\Lambda}_1 \leqq R_1 \leqq \Lambda_1, \tag{28}$$

und verbessert die Näherungen Λ aus der reduzierten Aufgabe in der Regel deutlich, wie mehrere Testbeispiele zeigen. Die numerische Auswertung des Rayleighquotienten (27) ist recht einfach, da der Zähler

$$Z_R(\boldsymbol{w}_i) = \boldsymbol{w}_i^T\boldsymbol{S}_w\boldsymbol{w}_i = \boldsymbol{m}_i^T\boldsymbol{G}^T\boldsymbol{S}_w^{-1}\boldsymbol{G}\boldsymbol{m}_i \tag{29}$$

bereits bei der Lösung der reduzierten Aufgabe auftaucht und dort zweckmäßig zu Eins normiert wird:

$$Z_R(\boldsymbol{w}_i) = 1 \rightarrow R_i = \frac{1}{\boldsymbol{w}_i^T\boldsymbol{H}\boldsymbol{w}_i}. \tag{30}$$

Damit verbleibt nur die Berechnung der quadratischen Form $\boldsymbol{w}^T\boldsymbol{H}\boldsymbol{w}$, die auf Elementebene durchgeführt werden kann, so daß die gesamte Systemmatrix $\boldsymbol{H}$ nicht explizit benötigt wird und somit kaum zusätzlicher Speicherplatz belegt wird.

Der Defekt der reinen Weggrößen-Eigenwertaufgabe (26) für die Näherungen $\boldsymbol{w}_i$, R_i (27) erlaubt eine Einschließung der Eigenwerte $\tilde{\Lambda}$. Da die Inverse $\boldsymbol{S}_w^{-1}$ der Steifigkeitsmatrix bereits vorliegt, geht man zweckmäßig von der transformierten Aufgabe

$$(\boldsymbol{H} - \gamma\boldsymbol{S}_w)\,\tilde{\boldsymbol{w}} = \boldsymbol{0}\,, \qquad \gamma = \tilde{\Lambda}^{-1}\,, \qquad \Gamma_i = R_i^{-1} \leqq \gamma_i\,, \tag{31}$$

aus und bestimmt zunächst obere Schranken γ_0 für γ, die dann in untere Schranken $\tilde{\Lambda}_u$ umgerechnet werden.

Das normierte Defektquadrat

$$\delta_i^2 = \frac{\boldsymbol{d}_i^T\boldsymbol{S}_w^{-1}\boldsymbol{d}_i}{\boldsymbol{w}_i^T\boldsymbol{S}_w\boldsymbol{w}_i}\,, \qquad \boldsymbol{d}_i = \boldsymbol{H}\boldsymbol{w}_i - \Gamma_i\boldsymbol{S}_w\boldsymbol{w}_i\,, \tag{32}$$

enthält bei der Normierung (29) mit (30) der Eigenvektoren $\boldsymbol{m}$ nur einen neu zu berechnenden Skalar,

$$\delta_i^2 = \boldsymbol{w}_i^T\boldsymbol{H}^T\boldsymbol{S}_w^{-1}\boldsymbol{H}\boldsymbol{w}_i - \Gamma_i^2\,, \qquad \boldsymbol{w}_i^T\boldsymbol{S}_w\boldsymbol{w}_i = 1\,, \tag{33}$$

und bestimmt nach dem Einschließungssatz von KRYLOV-BOGOLJUBOW eine obere Schranke γ_0 für γ beziehungsweise eine untere $\tilde{\Lambda}_u$ für $\tilde{\Lambda}$:

$$\gamma_0 = \Gamma + \delta \rightarrow \tilde{\Lambda}_u = \gamma_0^{-1}. \tag{34}$$

Solange die ersten Einschließungsbereiche, von $i = 1$ ausgehend, getrennt sind, gilt die verschärfte Einschließung

$$\gamma_0 = \Gamma + \delta^2 s^{-1} \rightarrow \tilde{\Lambda}_u = \gamma_0^{-1}, \qquad s_i = \Gamma_i - \gamma_{j0}, \tag{35}$$

nach TEMPLE mit den Bezeichnungen nach Bild 1.

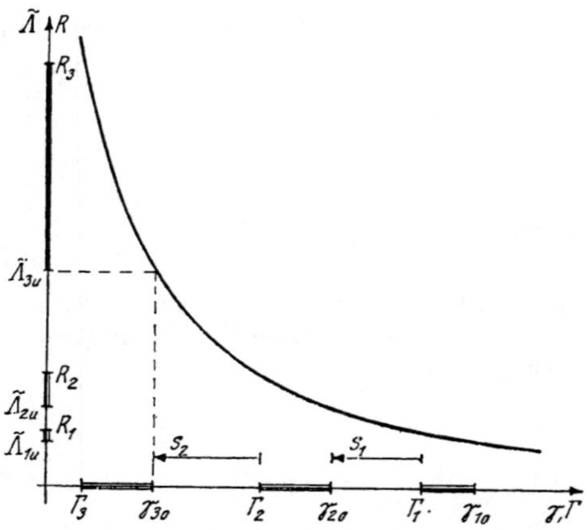

Bild 1. Einschließungsbereiche $\Gamma_i \leq \gamma_i \leq \gamma_{i0}$ nach KRYLOV für $\gamma = \tilde{\Lambda}^{-1}$ und Abstände s für die Verschärfung nach TEMPLE

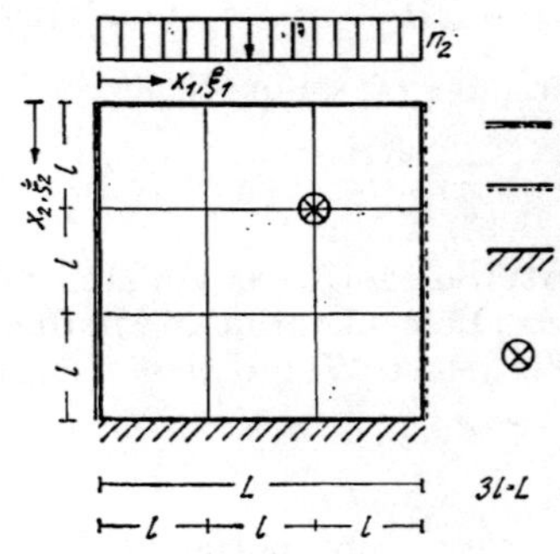

Bild 2. Berechnete Beispielplatte

5. Beispiel

Die Methode zur exakten Reduktion statischer Stabilitätsprobleme wurde anhand des BERNOULLIbalkens skizziert, um das Wesentliche des Verfahrens in möglichst knapper Form darzustellen.

Die besonderen Vorteile des neuen Verfahrens zeigen sich bei der Berechnung der Beullasten zweidimensionaler Kontinua. Hier wird das Beulen einer KIRCHHOFFschen Rechteckplatte nach Bild 2 unter Randspannungen n_2 vorgeführt.

Die Formänderungsenergie E_{FA} für ein Plattenelement der Dicke t mit den Kantenlängen a_1 in x_1-Richtung und a_2 in x_2-Richtung, jeweils bezogen auf eine charakteristische Länge l, und der POISSONziffer ν ist wohlbekannt Das aufgesetzte Dach $(\hat{z})$ kennzeichnet dimensionslose Zustandsgrößen.

$$x_i = a_i \xi_i, \qquad \alpha_i = l/a_i, \qquad 0 \leqq \xi_i \leqq 1, \qquad (\ )_{,i} = \partial(\ )/\partial\xi_i, \qquad \hat{w} = w/l,$$
$$2\hat{E}_{FA} = \int (\alpha_1^4 \hat{w}_{,11}^2 + 2\alpha_1^2\alpha_2^2 \hat{w}_{,11}\hat{w}_{,22} + \alpha_2^4 \hat{w}_{,22}^2 + 2(1-\nu)\,\alpha_1^2\alpha_2^2 \hat{w}_{,12}^2)\, d\xi_1\, d\xi_2. \tag{36}$$

Die Arbeit A_{ax} der konstanten Randspannungen n_2,

$$n_2 l^2/D = \lambda, \qquad D = \frac{Et^3}{12(1-\nu^2)}, \tag{37}$$

wird entsprechend der Aufteilung (11) als gemischtes Funktional in $\hat{w}$ und $\hat{m} = \lambda \hat{w}_{,2}$ angeschrieben:

$$\hat{A}_{ax} = -\int \alpha_2^2 \hat{w}_{,2} \hat{m}\, d\xi_1\, d\xi_2 + \tfrac{1}{2} \int \alpha_2^2 \lambda^{-1} \hat{m}^2\, d\xi_1\, d\xi_2, \qquad \hat{m} = \lambda \hat{w}_{,2}. \tag{38}$$

Die getrennte Variation des gemischten Funktionals $G = E_{FA} + A_{ax}$ als Summe aus (36) und (38) führt, wie im Abschnitt 2 für den BERNOULLIbalken dargelegt, auf die Interpretation $\hat{m} = \lambda \hat{w}_{,2}$ der Versatzmomente und die KIRCHHOFFsche Differentialgleichung der Plattenbeulung. Der stationäre Charakter des Funktionals G ist gewährleistet bezüglich all der Variationsfunktionen für w, welche für Kontinuität in den Randdurchsenkungen und -normalableitungen $w_{,n}$ sorgen. An die Variationen der Versatzmomente m werden keine Forderungen gestellt.

Als zulässige Ansatzfunktionen $\hat{w}(\xi_1, \xi_2) = \boldsymbol{w}^T \boldsymbol{\Phi}(\xi_1, \xi_2)$ entsprechend (17) für die Biegefläche eines Elementes mit den Eckpunkten i, j, k, l werden Produkte kubischer HERMITEpolynome benutzt, mit denen sich die Auswertung der Funktionalintegrale in (36) besonders einfach gestaltet.

$$\hat{w}(\xi_1, \xi_2) = \boldsymbol{w}^T \boldsymbol{\Phi}(\xi_1, \xi_2),$$
$$\boldsymbol{\Phi}(\xi_1, \xi_2) = \begin{bmatrix} h_1(\xi_2) & \boldsymbol{h}(\xi_1) \\ \cdot & \\ \cdot & \\ h_4(\xi_2) & \boldsymbol{h}(\xi_1) \end{bmatrix}, \qquad \boldsymbol{h}(\xi) = \begin{bmatrix} 1 & 0 & -3 & 2 \\ 0 & 1 & -2 & 1 \\ 0 & 0 & 3 & -2 \\ 0 & 0 & -1 & 1 \end{bmatrix} \begin{bmatrix} 1 \\ \xi \\ \xi^2 \\ \xi^3 \end{bmatrix}. \tag{39}$$

Jedem Knotenpunkt i sind dabei vier mechanisch deutbare Parameter zugeordnet:

$$\boldsymbol{w}_i^T = [\hat{w}\ \hat{w}_{,1}\ \hat{w}_{,2}\ \hat{w}_{,12}]_i. \tag{40}$$

Insgesamt wird das elastische Verhalten der Beispielplatte nach Berücksichtigung der geometrischen Randbedingungen durch $n_{w,\text{red}} = 64 - 29 = 35$ Weggrößen-Freiwerte angenähert.

Für die selbständige Approximation $\hat{m}(\xi_1, \xi_2) = \boldsymbol{m}^T\boldsymbol{\Psi}(\xi_1, \xi_2)$ der Versatzmomente m entsprechend (17) wird jedem Knoten i ein einziger Parameter $\hat{m}_i$ zugeordnet, der mit der Neigung $(\hat{w}_{,2})_i$ korrespondiert, was auf eine bilineare Interpolation im Element mit den Knoten i, j, k, l führt:

$$\hat{m}(\xi_1, \xi_2) = \boldsymbol{m}^T\boldsymbol{\Psi}, \qquad \boldsymbol{\Psi} = \begin{bmatrix} (1-\xi_2)\begin{bmatrix} 1-\xi_1 \\ \xi_1 \end{bmatrix} \\ \xi_2 \begin{bmatrix} 1-\xi_1 \\ \xi_1 \end{bmatrix} \end{bmatrix}, \qquad \boldsymbol{m} = \begin{bmatrix} \hat{m}_i \\ \hat{m}_j \\ \hat{m}_k \\ \hat{m}_l \end{bmatrix}. \tag{41}$$

Obwohl die Annäherung der Versatzmomente ohne Rücksicht auf Rand- und Übergangsbedingungen erfolgen kann ist es dennoch sinnvoll, für Stetigkeit längs der Elementfugen zu sorgen und die Knotenparameter $\hat{m}_i$ dort Null zu setzen, wo die zugeordneten Neigungen $(\hat{w}_{,2})_i$ verschwinden. Auf diese Weise bleiben $n_{m,\text{red}} = 9$ Freiwerte für die Approximation der Versatzmomente im Gesamtgebiet übrig.

Die finite Übersetzung des gemischten Funktionals $G = F_{F\lambda} + A_{ax}$ mit (36) und (38) führt auf eine Bilinearform (18) mit den Elementen (19) und dem Gradienten (20). Über die Elimination (21) der Weggrößenparameter erhält man schließlich das reduzierte Eigenwertproblem (22) der Ordnung $n_{m,\text{red}} = 9$.

Die bekannte numerische Gutmütigkeit des BERNOULLIschen Weggrößenfunktionals für bikubische Ansätze überträgt sich auf die Minimierung des dem Eigenwertproblem (22) zugeordneten RAYLEIGHquotienten (24). Die finite Übersetzung der Versatzmomente verwertet den Ansatz $\hat{m} = \boldsymbol{m}^T\boldsymbol{\Psi}$ optimal, da er ohne Aufrauhung durch Ableiten in das Funktional (24) eingeht.

Die Beulwerte λ werden einmal mit Hilfe des reduzierten Systems der Ordnung $n = 9$ entsprechend (22) angenähert (Ansatztyp A), wofür die ersten drei Eigenwerte Λ in der Tabelle 1 zu finden sind. Zum Vergleich wurde die Arbeit A_{ax} der Randspannungen auch als reines Weggrößenfunktional ausgewertet, was auf ein allgemeines Matrizeneigenwertproblem der Ordnung $n = 35$ mit den Näherungen $\tilde{\Lambda}$ für λ führt (Ansatztyp D).

Obere Schranken $\tilde{\Lambda}_0$ für diese Werte, die die Näherungen Λ aus der reduzierten Aufgabe noch verkleinern, also verbessern, erhält man mit Hilfe der RAYLEIGHquotienten R nach (27) mit den Eigenvektoren $\boldsymbol{m}_i$ zu Λ_i in der aufgeweiteten Form $\boldsymbol{w} = \boldsymbol{S}_w^{-1}\boldsymbol{G}\boldsymbol{m}$. (Ansatztyp B).

Ein Vergleich der auf $\tilde{\Lambda}_1$ ($n_{w,\text{red}} = 35$) bezogenen Eigenwertdifferenzen für die maßgebliche kleinste Beullast,

$$\Delta\Lambda = 100\,\frac{\tilde{\Lambda}_1 - \Lambda_1}{\tilde{\Lambda}_1} = -0{,}67\%\,, \qquad \Delta R = 100\,\frac{\tilde{\Lambda}_1 - R_1}{\tilde{\Lambda}_1} = -0{,}03\%\,, \tag{42}$$

einmal für die Näherungen Λ_1 ($n_{m,\text{red}} = 9$), zum anderen für R_1, unterstreicht zunächst die Brauchbarkeit der vorgeführten Reduktion und rechtfertigt den geringen Zusatzaufwand, nämlich proportional $n_{w,\text{red}}^2$, zur Auswertung des RAYLEIGHquotienten (30). Die unteren Schranken $\tilde{\Lambda}_u$ aus der verschärften Einschließung (35) nach TEMPLE runden die Beurteilung der Eigenwerte Λ ($n = 9$) gegenüber $\tilde{\Lambda}$ ($n = 35$) ab. (Ansatztyp C).

Tabelle 1. Näherungswerte für die ersten drei Beullasten $\lambda = n_2L^2/D$ aufgrund verschiedener Approximationen. $\nu = 0{,}3$

Typ der Approximation	$i = 1$	$i = 2$	$i = 3$
Typ A ($n = 9$); $\Lambda_i =$	13,0222	52,5734	82,0932
Typ B; $R_i = \tilde{\Lambda}_{io} =$	12,9392	51,8427	75,3427
Typ C; $\tilde{\Lambda}_{iu} =$	12,9345	51,5803	65,6554
Typ D ($n = 35$); $\tilde{\Lambda}_i =$	12,9355	51,7947	68,6590

Typ A: Eigenwerte Λ aus dem reduzierten System $n = 9$ mit A_{ax} als gemischtem Funktional

Typ B: RAYLEIGHquotient für $\tilde{\Lambda}$ mit den Eigenvektoren des reduzierten Systems ($n = 9$) als oberer Schranke für $\tilde{\Lambda}$

Typ C: Untere Schranke $\tilde{\Lambda}_u$ für $\tilde{\Lambda}$ aus verschärfter Einschließung nach TEMPLE mit den Eigenvektoren des reduzierten Systems

Typ D: Eigenwerte $\tilde{\Lambda}$ aus der finiten Übersetzung des reinen Weggrößenfunktionals mit $n = 35$

6. Abschließende Hinweise

Die im Beispiel deutlich gewordene merkliche Verbesserung der Approximation durch eine abschließende Berechnung der RAYLEIGHquotienten konnte bei allen Testbeispielen beobachtet werden, so daß diese Nachlaufrechnung auf jeden Fall zu empfehlen ist. Die Einschließung der Eigenwerte $\tilde{\Lambda}$ des zugeordneten reinen Weggrößenfunktionals nach KRYLOV mit der Verschärfung nach TEMPLE ist dagegen für die Rechenpraxis weniger aussagekräftig.

Über die Einführung unabhängig zu variierender „Versatzmomente" in die Arbeit der Axialkräfte beziehungsweise der Mittelflächenspannungen hinaus kann auch die Formänderungsenergie als stationäres Funktional mehrerer Zustandsgrößen formuliert werden. Allerdings erschwert die fehlende Extremaleigenschaft eine Beurteilung der Ergebnisse.

Literatur

1 Oden, J. T., Reddy, J. N., On dual-complementary variational principles in mathematical physics, Int. J. Engng. Sci. **12**, 1–29 (1974).
2 Tabarrok, B., Complementary energy method in elastodynamics, IUTAM Symposium "High speed computing of elastic structures", Les congrès et colloques de l'université de Liège **61**, 625–662 (1971).
3 Geradin, M., Computational efficiency of equilibrium models in eigenvalue analysis, IUTAM Symposium Liège **61**, 589–623 (1971).
4 Hughes, T. J. R., Reduction scheme for some structural eigenvalue problems by a variational theorem, Int. J. num. Meth. Engng. **10**, 845–852 (1976).
5 Hughes, T. J. R., A reduction scheme for problems of structural dynamics, Int. J. Solids Structutes **12**, 749–767 (1976).

Eingereicht am 11. 5. 1979

Anschrift: Dr.-Ing. Peter Ruge, Lehrstuhl für Mechanik und Festigkeitslehre im Mechanikzentrum der TU Braunschweig, Pockelstraße 4, D-3300 Braunschweig, BRD

ZAMM 61, 75–80 (1981)

V. Комкоv

Application of Invariant Variational Principles to the Optimal Design of a Column*)

Using Noether's technique we derive some invariants associated with the problems of optimization of design. While specifically a beam design problem is discussed here, the technique can be generalized in an obvious manner to thin plate, or shell problems. The specific problem discussed in this paper concerns the optimization of a column. Some invariants derived here as examples of the technique can be used to point out the basic difficulties discovered by much harder arguments in many recent papers. In particular, our method points out the existence of singularities which are associated with an optimal column design even in the absence of multiple eigenvalues in the Euler's eigenvalue problem, which corresponds to the buckling phenomenon.

Unter Verwendung des Verfahrens von Noether werden einige Invarianten hergeleitet, die bei der Optimierung von Strukturen auftreten. Obwohl speziell ein Balkenoptimierungsproblem diskutiert wird, kann das Verfahren auf dünne Platten oder Schalen verallgemeinert werden. Das speziell in dieser Arbeit behandelte Problem bezieht sich auf die Optimierung eines Stabes. Einige hier als Beispiele des Verfahrens hergeleitete Invarianten können dazu benutzt werden, die prinzipiellen Schwierigkeiten aufzuzeigen, die in vielen neuen Arbeiten erst durch viel schwerfälligere Argumente sichtbar werden. Insbesondere zeigt unsere Methode die Existenz von Singularitäten auf, welche mit dem optimalen Entwurf eines Stabes auch beim Fehlen von mehrfachen Eigenwerten beim Eulerschen Eigenwertproblem entsprechend der Stabknickung verbunden sind.

Используя метод Нетера выводятся некоторые инварианты, появляющиеся при оптимировании структур. Хотя здесь обсуждается вопрос оптимирования балки, этот метод можно обобщить на тонкие пластины и оболочки. Обсуждаемый в данной работе вопрос касается оптимирования стержня. Некоторые полученные здесь при демонстрации метода инварианты могут быть использованы для выявления принципиальных трудностей, которые во многих новых работах становятся очевидными только посредством более тяжеловесных аргументов. Наш метод показывает, в частности, существование сингулярностей, связанных с оптимальной схемой стержня также при отсутствии многократных собственных значений в задаче о собственных значениях Эйлера, что соответствует явлению продольного изгиба.

1. Introduction

A classical version of Noether's theorem concerns the invariance of the Lagrangian of motion with respect to a k-parameter continuous group of transformations. In the original article of E. Noether [18], the Lagrangian of motion is of the simple form $L = L(x, \dot{x}, t)$, however this is not a serious restriction and much more complex functionals can be treated by Noether's technique. See for example [1, 2].

The work of Noether considered the transformation properties of the differential equations arising from a variational formulation under a k-parameter group of transformations. The Noether theorem (theorems) establish the following facts. If the functional J is invariant under a k-parameter continuous group of transformations of the variables, then k-identities can be derived between the variational equations (i.e. Euler-Lagrange equations in the case considered by Noether) and certain quantities which can be written in a divergence form.

If the parameters are replaced by k-arbitrary functions and their derivatives, then appropriate relations can be established between the variational equations and their derivatives. The problem discussed in this paper is in many ways analogous to the minimization of the Lagrangian functional which was considered by Noether.

Suppose that we consider a functional $J(y)$ of the form $J(y) = L(x, y^i(x), y^{i'}(x))\,\mathrm{d}x$ and a k-parameter group of transformations with parameters ε^i, $i = 1, 2, \dots, k$. If we demand that $J(y)$ is divergence invariant (i.e. we require that

$$\left(L(x, y^i(x), y^{i'}(x)) - \overline{L}\left(\bar{x}, \bar{y}^i(\bar{x}), \frac{\mathrm{d}\bar{y}^i}{\mathrm{d}x}\right)\frac{\mathrm{d}\bar{x}}{\mathrm{d}x}\right) = \varepsilon^s \frac{\mathrm{d}}{\mathrm{d}x}(\Phi_s) + o(\varepsilon)$$

for some C^1 functions Φ_s), Noether's theorem requires the functions Φ_s to satisfy k-separate identities of the form:

$$\frac{\partial L}{\partial x}T_s + \frac{\partial L}{\partial y^i}\xi_s^i + \frac{\partial L}{\partial y^{i'}}\left(\frac{\mathrm{d}\xi_s^i}{\mathrm{d}x} - y^{i'}\frac{\mathrm{d}T_s}{\mathrm{d}x}\right) + L\frac{\partial T_s}{\mathrm{d}x} = \frac{\mathrm{d}\Phi}{\mathrm{d}x}, \qquad s = 1, 2, \dots, k,$$

where ξ_s^i, T_s are the infinitesimal generators of the group of transformations: $\bar{x} = \varphi(x, y, \varepsilon)$, $\bar{y}^i = \psi^i(x, y, \varepsilon)$, which are defined by the relation $\xi_s^i = \dfrac{\partial\psi^i(x, y, 0)}{\partial\varepsilon_s}$, $T_s = \dfrac{\partial\varphi(x, y, 0)}{\partial\varepsilon_s}$.

To assume the invariance of the cost functional (or "action") integral, one needs to specify the invariance of the k-integrals $\left(\varepsilon_s \dfrac{\mathrm{d}\Phi_s}{\mathrm{d}x}\right)\mathrm{d}x$ in the sense that $\left(\varepsilon_s \dfrac{\mathrm{d}\Phi_s}{\mathrm{d}x}\right)\mathrm{d}x = o(\varepsilon)$, $s = 1, 2, \dots, k$. In particular if $\dfrac{\mathrm{d}\Phi_s}{\mathrm{d}x} \equiv 0$, then invariance of the functional $J = L(x, y, y')\,\mathrm{d}x$ follows trivially from Noether's theorem. This remark will be used repeatedly in computing the invariants of a specific cost functional λ of which we shall regard as "our Lagrangian".

*) Research supported by U.S. National Science Foundation, Grant ENG 77-19967.

In the present paper we use NOETHER's ideas to derive various integral invariants associated with different classes of optimization problems for elastic systems. Our goal posed in this paper is to derive physically significant invariants which are directly related to the problem of optimization of elastic beams. We shall consider in particular the problems of optimization of the cross-sectional area of a column against buckling, and a closely related problem of optimization of a vibrating beam.

Let us consider the linear (approximate) differential equation representing EULER's buckling problem

$$(EI(x)\, y'')'' + P_{cr} y'' = 0\,, \qquad y(0) = y(l) = 0\,, \tag{1.1}$$

which can be derived from the energy equality

$$\tfrac{1}{2} \int_0^l (EIy'')^2\, dx = \tfrac{1}{2} \int_0^l P_{cr} (y')^2\, dx\,. \tag{1.2}$$

We regard, however, the cross-sectional area of the beam as a variable $A = A(x)$, but assume the existence of a functional relation $I(x) = \tilde{I}(A(x))$ between the moment of inertia of the cross-sectional area $I(x)$ and its magnitude $A(x)$. A well explored class of optimization problems consists of either maximizing the first eigenvalue P_{cr_1} subject to the constraint

$$W = \int_0^l A(x)\, dx = \text{constant}\,,$$

or of minimizing W subject to the constraint $P_{cr_1} \geqq K > 0$, where K is a given value. See [3, 4, 5, 6, 7, 8, 9, 10, 11] for a sampling of relevant literature. Reference [3] is of particular importance since it is largely responsible for many current trends. The beam optimization problem consists in choosing $A(x) \geqq 0$, $0 \leqq x \leqq l$, such that

$$\min_{y \in Y} \left\{ \int_0^l E\tilde{I}(A(x))\, (y'')^2\, dx \Big/ \int_0^l (y')^2\, dx \right\} \geqq K\,,$$

and

$$\int_0^l A(x)\, dx = \text{minimum}\,. \tag{1.3}$$

The class of admissible displacement functions Y consists of $H_0^2[0, l]$ functions which satisfy assigned boundary conditions at $x = 0$ and at $x = l$, such as

$$y(0) = y(l) = 0\,, \qquad I(0)\, y'' = I(l)\, y''(l) = 0 \tag{1.4}$$

(a free support at each end).

We postulate that the solution $A(x)$ must be a non-negative continuous function whose derivative dA/dx exists almost everywhere, such that

$$\int_0^l (EA(x))^{-1}\, dx < \infty\,. \tag{1.5}$$

We postulate $A(x) \in H_0^1[0, l]$, and the condition (1.5). For the physical reasoning justifying the necessity of the constraint (1.5) see [11].

The minimization problem as stated above in (1.3) is associated with stationary behavior of the functional

$$\lambda(A, y) = \mu_0 \int_0^l A(x)\, dx - \frac{1}{2}\beta \left(\int_0^l E\tilde{I}(A(x))\, (y'')^2\, dx - K \right) - \frac{\mu}{2} \left(\int_0^l (y')^2\, dx - 1 \right) = \int_0^l \lambda(0)\, dx\,, \tag{1.6}$$

where β, μ are Lagrangian multipliers. More specifically, if $\lambda(A)$ is a FRÉCHET differentiable functional of the vector $A \in H_1[0, l]$, then $\partial\lambda/\partial A = 0$ is a necessary condition for the optimality of $\lambda\colon H^1 \to R$, and the vanishing of $\partial\lambda/\partial y$ is in fact the statement that the equation

$$(EIy'')'' + P_{cr} y'' = 0 \tag{1.1}$$

is satisfied, combined with the inequality

$$P_{cr} \geqq K$$

(which, by the way, can be replaced by $P_{cr} = K$, without any loss of generality).

2. A One-Dimensional Version of Noether's Theorem

In this article we consider the one-dimensional version of NOETHER's theorem concerning a k-parameter continuous group of transformations. We shall closely follow the original arguments of E. NOETHER [18], concerning the properties of the functional $\lambda(A, y)$ given by the relation (1.6). Following the classical formulation of EMMY NOETHER, we introduce a k-parameter $(\varepsilon_1, \varepsilon_2, \ldots, \varepsilon_k)$ continuous group of transformations G_k:

$$\bar{x} = \varphi(A, x, y, \varepsilon)\,, \qquad \bar{y} = \psi(A, x, y, \varepsilon)\,, \qquad \bar{A} = \nu(A, x, y, \varepsilon)\,, \tag{2.1}$$

such that

$$\bar{x} = \varphi(A, x, y, 0) = x\,, \qquad \bar{y} = \psi(A, x, y, 0) = y\,, \qquad \bar{A} = \nu(A, x, y, 0) = A\,. \tag{2.2}$$

The transformations $(x, y, A) \xrightarrow{\varepsilon} (\bar{x}, \bar{y}, \bar{A})$ do form a group and must have a uniquely defined inverse for each admissible value of the parameter vector ε. This presents a problem of choosing the class of transformations which is "naturally" suited to our physical problem. Since the Noether invariants depend on our choice of the transformation, the depth of the results is directly related to our ability of choosing the group G.

The infinitesimal generators of the group G are given by

$$\tau_i = \frac{\partial \varphi}{\partial \varepsilon_i}\bigg|_{\varepsilon=0}, \qquad \xi_i = \frac{\partial \psi}{\partial \varepsilon_i}\bigg|_{\varepsilon=0}, \qquad \zeta_i = \frac{\partial \nu}{\partial \varepsilon_i}\bigg|_{\varepsilon=0}, \qquad i = 1, 2, \dots, k .$$

We shall choose some k-parameter groups G_k of transformations, mapping

$$x \in [0, l] \text{ into } \bar{x}, \qquad y \in H_0^2[0, l] \text{ into } \bar{y} \in H_0^2[0, l],$$

$$A > 0, \qquad A \in H_0^1[0, l] \text{ into } \bar{A} \in H_0^1[0, l].$$

We comment that in general $\bar{x}$, $\bar{y}$, $\bar{A}$ do not satisfy physical requirements imposed on x, y, and A. For example $\bar{A}$ may become negative, $\bar{y}(\bar{x})$ may exhibit unrealistic shapes etc. This is immaterial, since the transformation applied to x, y, A serves only as a means for deriving some properties of the functional (1.6) and thereby of the whole class of buckling or vibration problems. We do not have to regard it as a physically admissible transformation. The functional λ is invariant under the action of the group G_k if for each $i = 1, 2, \dots, k$, the equation

$$\frac{\partial \lambda}{\partial x}\tau_i + \frac{\partial \lambda}{\partial A}\eta_i + \frac{\partial \lambda}{\partial y}\xi_i + \frac{\partial \lambda}{\partial y_x}\left(\frac{\mathrm{d}\xi_i}{\mathrm{d}x} - y_x\frac{\mathrm{d}\tau_i}{\mathrm{d}x}\right) + \frac{\partial \lambda}{\partial y_{xx}}\left(\frac{\mathrm{d}^2\xi_i}{\mathrm{d}x^2} - 2y_{xx}\frac{\mathrm{d}\tau_i}{\mathrm{d}x} - y_x\frac{\mathrm{d}^2\tau_i}{\mathrm{d}x^2}\right) + \lambda\frac{\mathrm{d}\tau_i}{\mathrm{d}x} = 0 \tag{2.3}$$

is satisfied.

Since in the case of beam optimization represented by functional (1.6) λ is not an explicit function of x or of y, the equation (2.3) may be simplified into the form

$$\frac{\partial \lambda}{\partial A}\eta_i + \frac{\partial \lambda}{\partial y_x}\left(\frac{\mathrm{d}\xi_i}{\mathrm{d}x} - y_x\frac{\mathrm{d}\tau_i}{\mathrm{d}x}\right) + \frac{\partial \lambda}{\partial y_{xx}}\left(\frac{\mathrm{d}^2\xi_i}{\mathrm{d}x^2} - 2y_{xx}\frac{\mathrm{d}\tau_i}{\mathrm{d}x} - y_x\frac{\mathrm{d}^2\tau_i}{\mathrm{d}x^2}\right) + \lambda\frac{\mathrm{d}\tau_i}{\mathrm{d}x} = 0, \tag{2.3a}$$

where $\mathrm{d}/\mathrm{d}x$ denotes a total derivative; for example

$$\frac{\mathrm{d}\xi_i}{\mathrm{d}x} \equiv \frac{\partial \xi_i}{\partial x} + \frac{\partial \xi_i}{\partial y}y_x + \frac{\partial \xi_i}{\partial y_x}y_{xx} + \frac{\partial \xi_i}{\partial y_{xx}}y_{xxx} + \frac{\partial \xi_i}{\partial A}\frac{\partial A}{\partial x}.$$

In the classical application of Noether's theory, the transformation consisting of time translation gives the invariance of the Hamiltonian, if the Lagrangian does not implicitly depend on time. Regarding our functional λ as "the Lagrangian", and noting that λ does not explicitly depend on x, we obtain an invariant of the system by setting

$$\bar{x} = x + \varepsilon_1, \qquad \bar{A} = A, \qquad \bar{y} = y.$$

Then $\tau_1 = 1$ and the remaining infinitesimal generators of G: τ_i, ξ_i, η_i are all equal to zero. Hence, we have

$$\lambda - y_x\frac{\partial \lambda}{\partial y_x} + y_x\frac{\mathrm{d}}{\mathrm{d}x}\left(\frac{\partial \lambda}{\partial y_{xx}}\right) - y_{xx}\frac{\partial \lambda}{\partial y_{xx}} = \text{constant}, \tag{2.4}$$

or if we use (1.6)

$$\frac{\mathrm{d}}{\mathrm{d}x}\left[\lambda + \mu(y')^2 - \beta y'\frac{\mathrm{d}}{\mathrm{d}x}(EIy'') + \beta EI(y'')^2\right] = 0, \tag{2.4a}$$

and

$$\frac{\partial \lambda}{\partial A} = 0.$$

These relations are valid for all $x \in [0, l]$. The term

$$\frac{\partial \lambda}{\partial y_x} - \frac{\mathrm{d}}{\mathrm{d}x}\left(\frac{\partial \lambda}{\partial y_{xx}}\right) = p \tag{2.4b}$$

is analogous to the canonical momentum for a Lagrangian which depends on both velocity and acceleration (See H. S. P. Grässer [12].).

The expression (2.4) is analogous to the Hamiltonian

$$H = L - \dot{x}p - \ddot{x}\frac{\partial L}{\partial \ddot{x}}, \quad \text{where} \quad p = \frac{\partial L}{\partial \dot{x}} - \frac{\mathrm{d}}{\mathrm{d}t}\left(\frac{\partial L}{\partial \ddot{x}}\right).$$

This Hamiltonian is constant along any trajectory of motion if the Lagrangian L does not explicitly depend on time. We observe also that λ does not explicitly depend on $y(x)$.

Introducing the transformation

$$\bar{x} = x, \qquad \bar{A} = A, \qquad \bar{y} = \varepsilon + y,$$

we obtain the analogue of a well-known invariant of classical mechanics

$$\frac{\partial\lambda}{\partial y_x} - \frac{\mathrm{d}}{\mathrm{d}x}\left(\frac{\partial\lambda}{\partial y_{xx}}\right) = \text{constant}. \tag{2.5}$$

Inserting (2.5) into (1.6) we have

$$-\mu y' + \beta\frac{\mathrm{d}}{\mathrm{d}x}(EI(A)\,y'') = \text{constant}. \tag{2.5a}$$

Inserting this expression into (2.4), we obtain

$$\lambda + y'\ (\text{constant}) - y_{xx}\frac{\partial\lambda}{\partial y_{xx}} = \text{constant},$$

or for some constant α

$$\lambda + \alpha y' + \beta EI(y'')^2 = \text{constant}, \tag{2.5b}$$

and

$$\frac{\partial\lambda}{\partial A} = 0, \tag{2.5c}$$

valid for all $x \in [0, l]$.

We are ready to experiment with more complex identities. We consider the transformation (2.1)—(2.2) of the form

$$\bar{x} = x + \varepsilon_1 x(l - x), \qquad \bar{y} = x + \varepsilon_2 y, \qquad \bar{A} = A + \varepsilon_3 y + \varepsilon_4 A, \tag{2.6}$$

restricting the values of x to the interval $0 \leqq x \leqq l$, and $A \geqq 0$. The mapping $(x, y, A) \to (\bar{x}, \bar{y}, \bar{A})$ is one to one and has an inverse, if x is restricted to the interval $[0, l]$.

We compute the infinitesimal generators of the corresponding group G_4,

$$\begin{gathered}\tau_1 = x(l - x), \quad \tau_2 = \tau_3 = \tau_4 = 0; \quad \xi_1 = 0, \quad \xi_2 = y, \quad \xi_3 = \xi_4 = 0;\\ \nu_1 = \nu_2 = 0, \quad \nu_3 = y, \quad \nu_4 = A.\end{gathered} \tag{2.7}$$

The functional λ is invariant under this 4-parameter family of transformations if

$$\frac{\partial\lambda}{\partial x}\tau_i + \frac{\partial\lambda}{\partial A}\nu_i + \frac{\partial\lambda}{\partial y_x}\left(\frac{\mathrm{d}\xi_i}{\mathrm{d}x} - y_x\frac{\mathrm{d}\tau_i}{\mathrm{d}x}\right) + \frac{\partial\lambda}{\partial y}\xi_i + \frac{\partial\lambda}{\partial y_{xx}}\left(\frac{\mathrm{d}^2\xi_i}{\mathrm{d}x^2} - 2y_{xx}\frac{\mathrm{d}\tau_i}{\mathrm{d}x} - y_x\frac{\mathrm{d}^2\tau_i}{\mathrm{d}x^2}\right) + \lambda\frac{\mathrm{d}\tau_i}{\mathrm{d}x} = 0. \tag{2.8}$$

The system (2.8) after substituting the identities (2.7) becomes

$$\frac{\partial\lambda}{\partial x}x(l - x) - \frac{\partial\lambda}{\partial y_x}(y_x \cdot (l - 2x)) - \frac{\partial\lambda}{\partial y_{xx}}(y_{xx}(l - 2x) - 2y_x) + \lambda(l - 2x) = 0, \tag{2.9a}$$

$$\frac{\partial\lambda}{\partial y_x}(y_x) + \frac{\partial\lambda}{\partial y_{xx}}(y_{xx}) + \frac{\partial\lambda}{\partial y}y = 0, \qquad \frac{\partial\lambda}{\partial A}y = 0, \qquad \frac{\partial\lambda}{\partial A}A = 0. \tag{2.9 b, c, d}$$

Equations (2.9c) and (2.9d) state that

$$\frac{\partial\lambda}{\partial A} = 0 \tag{2.10}$$

except at points where $(y = 0 \wedge A = 0)$. Observing that $\frac{\partial\lambda}{\partial x} = \frac{\partial\lambda}{\partial y} = 0$, and substituting (2.9b) into (2.9c), we obtain the following system

$$(y \cdot A)\frac{\partial\lambda}{\partial A} = 0, \qquad \frac{\partial\lambda}{\partial x} = \frac{\partial\lambda}{\partial y} = 0, \tag{2.11), (2.12}$$

$$\frac{\partial\lambda}{\partial y_x}y_x + \frac{\partial\lambda}{\partial y_{xx}}y_{xx} = 0, \qquad -2y_x\frac{\partial\lambda}{\partial y_{xx}} + (l - 2x) = 0, \tag{2.13), (2.14}$$

expressing the necessary conditions for the invariance of λ. Moreover the following "conservation laws" are immediately derived.

Along the length of the column the following energy density forms are constant:

$$C_i = \lambda\tau_i + \left(\frac{\partial\lambda}{\partial y_x} - \frac{\mathrm{d}}{\mathrm{d}x}\left(\frac{\partial\lambda}{\partial y_{xx}}\right)\right)(\xi_i - y_x\tau_i) + \frac{\partial\lambda}{\partial y_{xx}}\frac{\mathrm{d}}{\mathrm{d}x}(\xi_i - y_x\tau_i). \tag{2.15}$$

Writing these identities in detail, we obtain

$$C_1 = \lambda \cdot x(l - x) - \left(\frac{\partial\lambda}{\partial y_x} - \frac{\mathrm{d}}{\mathrm{d}x}\left(\frac{\partial\lambda}{\partial y_{xx}}\right)\right)(x(l - x)\,y_x) - \frac{\partial\lambda}{\partial y_{xx}}\frac{\mathrm{d}}{\mathrm{d}x}(x(l - x)\,y_x), \tag{2.15a}$$

$$C_2 = \left(\frac{\partial\lambda}{\partial y_x} - \frac{\mathrm{d}}{\mathrm{d}x}\left(\frac{\partial\lambda}{\partial y_{xx}}\right)\right)y + \frac{\partial\lambda}{\partial y_{xx}}y_x, \qquad C_3 = C_4 \equiv 0. \tag{2.15b}$$

Substituting (1.6) into (2.15b), we obtain

$$C_2 = \left(-\mu y_x + \beta \frac{\mathrm{d}}{\mathrm{d}x}(EIy_{xx})\right) y - \beta EIy_{xx}y_x = -\mu y y_x + \beta y \frac{\mathrm{d}}{\mathrm{d}x}(EI(x)\, y_{xx}) - \beta y_x\, EIy_{xx}\,,$$

which is a constant of the deformation, and

$$\frac{\mathrm{d}C_2}{\mathrm{d}x} \equiv 0\,, \qquad \forall x \in [0, l]\,.$$

A numerical procedure.

The required energy density function is given by

$$\lambda(x) = \mu_0 A(x) - \frac{1}{2}\,\beta EI(A(x))\, y_{xx}^2 - \frac{\mu}{2}\, y_x^2 - k\,, \tag{2.17}$$

where k is a constant, $k = \dfrac{\beta K + \mu}{2l}$.

Using the relations (2.5a) and (2.5c) we can proceed as follows. The equation (2.5a) states

$$\mu_0 = \frac{1}{2}\beta EI'(A)\, y_{xx}^2\,, \quad \left({}' = \frac{\mathrm{d}}{\mathrm{d}A}\right). \tag{2.18}$$

For simplicity's sake let us assume $I(A) = CA^2$, so that $I'(A) \cong A$, the constant factor being absorbed by β, and choose $\mu_0 \equiv 1$. Then

$$y_{xx}^2 = 2/(\beta EA)\,. \tag{2.18a}$$

Using the invariant C_2 given by equation (2.15), we have

$$y_x(-\mu y - \beta EIy_{xx}) + \beta y \frac{\mathrm{d}}{\mathrm{d}x}(EIy_{xx}) = C_2\,.$$

Substituting $I \sim A^2$ and using the equation (2.5a) i.e.

$$-\mu y_x + \beta \frac{\mathrm{d}}{\mathrm{d}x}(EIy_{xx}) = C_3$$

we obtain

$$C_3 y - \beta EA^2 y_x y_{xx} = C_2$$

or the equation

$$y_x + kA^{-3/2} y = CA^{-3/2} \tag{2.19}$$

where k and C are constants, with initial condition $y(0) = 0$.

We can solve this equation in a closed form

$$y(x) = \exp\left(-\int_0^x kA^{-3/2}(\xi)\,\mathrm{d}\xi\right)\left\{\int_0^x \left(CA^{-3/2}(\xi)\left[\exp\left(\int_0^x kA^{-3/2}(\tau)\,\mathrm{d}\tau\right)\right]\right)\mathrm{d}\xi\right\}.$$

Because of the expected symmetry $\int_0^x$ can be replaced by $\int_l^{l-x}$ without altering the value of $|y(x)|$. Hence, if $A(x)$ is known at any point $x \in [0, l]$, $y(x)$ can be calculated after a choice of constants k and C. If $y(x) \neq 0$, y_x can be found from the relation (2.19).

However, if $y(x)$ is known at three consecutive points of a grid imposed on the beam, then $A(x)$ can be calculated from the relation (2.18). This suggests an iterative scheme which would completely avoid the usual techniques used in optimization and rely on the use of integral invariants to implement the numerical procedure.

3. Comments

Since we are solving in a systematic manner a first-order differential equation (2.19) only one boundary condition can be assigned to it, unless the solution has discontinuities in y_x, or else if by some extreme stroke of luck the boundary conditions assigned to $y(x)$, i.e. $y(0) = y(l) = 0$ just happen to be satisfied. Equation (2.18a) clearly indicates that at such points of discontinuity of y_x the cross-sectional area $A(x)$ must be equal to zero, that is a hinge develops at such singular points of the optimal design.

The author has come to the conclusion that while such singularities are mathematically justified, the fault lies with the mathematical description of the physical phenomena.

This aspect is discussed in a separate paper co-authored by E. J. Haug [11].

References

1 Anderson, D., Noether's theorem in general mechanics, J. Phys. A., **6**, (1973), 299–305.
2 Drobot, S. and Rybarski, A., A variational principle in hydromechanics, Arch. Rat. Mech. Anal, **2**, No. 5, (1958–1959), 393–410.
3 Keller, J. B., The shape of the strongest column, Arch. Rat. Mech. Anal., **5**, (1960), 275–285.
4 Tadjbakhsh, I. and Keller, J. B., Strongest columns and isoperimetric inequalities for eigenvalues, J. Appl. Mech., **9**, (1962), 159–164.
5 Keller, J. B. and Niordson, F. I., The tallest column, J. Math. Mech., **16**, (1966), 433–446.
6 Taylor, J. E., The strongest column: An energy approach, J. Appl. Mech., **34**, (1967), 486–492.
7 Olhoff, N. and Rasmussen, S. H., On single and bimodal buckling loads of clamped columns, Int. J. Solids Struct., **13**, (1977), 605–614.
8 Olhoff, N., Optimal design against structural vibration and instability, Ph. D. Thesis, Technical University of Denmark, Lyngby, Denmark, 1978.
9 Olhoff, N. and Niordson, F. I., Some problems concerning singularities of optimal beams and columns, ZAMM, **59**, (1979) T16–T26.
10 Masur, E. F., Optimality in the presence of discreteness and discontinuity, Proc. IUTAM, Symposium on optimization in structural design, Warsaw 1974, A. Sawczuk and Z. Mróz, editors, Springer-Verlag, Berlin, 1975.
11 Haug, E. J. and Komkov, V., The optimal shape of a column, to appear.
12 Grässer, H. S. P., A monograph on the general theory of second-order parameter invariant problems in the calculus of variations, Math. Commun. of Univ. of South Africa, Pretoria, 1967.
13 Komkov, V., A dual formulation of Noether's theorem with applications to continuum mechanics, to appear in J. Math. Anal. Applic. (1981).
14 Logan, J. D., Invariant variational principles, Math. in Science and Engineering Series, vol. **138**, Academic Press, New York, 1977.
15 Edelen, D. G. B., Nonlocal variations and local invariance of fields, American Elsevier, New York, 1969.
16 Eisenhart, L. P., Continuous groups of transformations, Princeton Univ. Press, Princeton, N.J., 1933.
17 Rund, H., A direct approach to Noether's theorem in the calculus of variations, Utilitas Math., **2**, (1972), 205–214.
18 Noether, E., Invariante Variationsprobleme, Nachr. Akad. Wiss. Göttingen, Math. Phys. Kl. II, (1918), 235–257.

Eingereicht am 29. 10. 1979, in revidierter Fassung am 20. 6. 1980

Anschrift: Prof. Vadim Komkov, College of Engineering, The University of Iowa, Iowa City, Iowa 52242, and The West Virginia University, Morgantown W. V. 26506, USA.

ZAMM 61, 81–87 (1981)

T. HATA

Thermoelastic Problem for a Griffith Crack in a Plate whose Shear Modulus is an Exponential Function of the Temperature

In this paper, we examine an analysis of thermal stresses taking into account the temperature-dependent properties, in which the variation of the shear modulus with temperature assumed to be an exponential function. Using the perturbation method, general equations for the displacements are found. The equations can be solved by using four displacement functions. Following this method, a solution is derived for calculating the stress intensity factor for a Griffith crack in a temperature-sensitive plate under a linear temperature distribution, where we assume that the temperature field is not disturbed by the crack. (Under these conditions, if a plate is temperature-insensitive, the stress intensity factor for the crack is zero.) In this case, whether the stress intensity factor is zero or not depends only on the temperature variation of the coefficient of thermal expansion. A numerical example is given for a plate made of steel. Results show the effect of temperature-dependent properties clearly.

In dieser Arbeit untersuchen wir Wärmespannungen, indem wir temperaturabhängige Eigenschaften berücksichtigen, wobei die Abhängigkeit des Schubmoduls von der Temperatur als eine Exponentialfunktion angenommen wird. Mittels der Störungsmethode werden allgemeine Gleichungen für die Verschiebungen gefunden. Unter Benutzung von vier Verschiebungsfunktionen können die Gleichungen gelöst werden. Entsprechend diesem Vorgehen wird eine Lösung zur Berechnung des Spannungsintensitätsfaktors für einen Griffith-Riß in einer temperaturempfindlichen Platte bei linearer Temperaturverteilung hergeleitet, wobei angenommen wird, daß das Temperaturfeld durch den Riß nicht beeinflußt wird. Ein numerisches Beispiel wird für eine Stahlplatte ausgeführt. Die Ergebnisse zeigen klar den durch die temperaturabhängigen Eigenschaften ausgelösten Effekt.

В данной работе изучаются тепловые напряжения при учёте зависящих от температуры свойств, причём, зависимость модуля сдвига от температуры предполагается в виде экспоненциальной функции. С помощью метода возмущений получены общие уравнения для перемещений. Уравнения могут быть разрешены при использовании четырёх функций перемещения. Сообразно изложенному методу найдено решение для рассчёта фактора интенсивности напряжения для трещины Гриффита в пластине, воспринимающей температурные изменения, при линейном распределении температуры в предположении, что трещина не влияет на температурное поле. Приведён численный пример для стальной пластины. Результаты ясно показывают эффекты, вызванные зависящими от температуры свойствами.

Introduction

Progress in the investigation of thermoelastic problems for temperature-sensitive materials has been made by TROSTEL [1], who applied the perturbation method to the problems of a cylinder and a beam. In this analysis, the variation of the coefficient of shear modulus with temperature is assumed to be a quadratic function of the temperature. The characteristics of this perturbation method for solving displacement equations are such that it can be applied to any kind of temperature-dependent thermoelastic problem with variable coefficients (except POISSON's ratio). But because of mathematical difficulties in solving these equations, the analyses of thermal-stress problems are restricted to simpler configurations, such as beams, polar-symmetric solids, or cylinders.

Recently, NYUKO, TAKEUTI and NODA [2] analyzed the temperature-dependent thermoelastic problem in a composite sphere. In the study, the variation of the coefficient of shear modulus with temperature in a sphere is assumed to be an exponential function of temperature.

In connection with the estimation of the strength of structural components at high temperature, it is important to discuss the effect of temperature-dependent properties on the thermal stress around a crack. But there seems to be no works concerned about the thermal stress intensity factor around a crack in the medium with temperature-dependent properties excepting the author's work [3].

In this paper, we analyze the stress intensity factor for a GRIFFITH crack in a plate with temperature-dependent properties under a linear temperature distribution, where we assume that the crack does not disturb the temperature field and the variation of the shear modulus with temperature is an exponential function. In this case, if the plate is insensitive, the stress intensity factor for the crack is zero. Then, it should be noted that the stress intensity factor for the crack has a value only for a sensitive plate.

The present paper also is an investigation of the general method for obtaining the solution of the thermoelastic problem for temperature-sensitive materials. Adopting the procedure of the perturbation method, we find that the solution of this problem may be obtained by using four displacement functions. A numerical example is given for a GRIFFITH crack in a plate made of steel. The results show the effect of temperature dependency of the material properties clearly.

Formulation of the Problem

By denoting the rectangular coordinates as x_i, and the time as t, we take the temperature distribution throughout the solid to be $T(x_i, t)$. Solution of the problem requires the determination of the stress components $\sigma_{ij}(x_i, t)$ and the strain components $e_{ij}(x_i, t)$ in the region Σ with boundary Γ. The stress-strain relations may be expressed as follows:

$$\sigma_{ij} = \lambda_T \delta_{ij} e + 2\mu_T e_{ij} - (3\lambda_T + 2\mu_T)\, \delta_{ij} \int_0^T \alpha_T(\eta)\, d\eta \tag{1}$$

where $e = e_{ii}$, λ_T, μ_T are LAME's constants, and δ_{ij} is KRONECKER's delta.

The equilibrium condition may be written as

$$\sigma_{ij,j} = 0\,. \tag{2}$$

The strain-displacement relations are

$$e_{ij} = (u_{i,j} + u_{j,i})/2 \tag{3}$$

where the u_i designate the displacements.

Making use of the postulate of thermal sensitiveness of the plate, we assume that the coefficient α_T is a function of temperature. But we assume that POISSON's ratio ν is constant in the range of temperature considered. YOUNG's modulus and the shear modulus μ_T have to be assumed temperature-dependent.

Substitution of Eqs. (1) and (3) into Eq. (2) yields the displacement equations of equilibrium

$$\mu_T\, \Delta u_i + (\lambda_T + \mu_T)\, e_{,i} + \lambda_{T,i} e + 2\mu_{T,j} e_{ij} = (3\lambda_T + 2\mu_T)_{,i} \int_0^T \alpha_T(\eta)\, d\eta + (3\lambda_T + 2\mu_T) \left(\int_0^T \alpha_T(\eta)\, d\eta\right)_{,i} \tag{4}$$

where $\Delta = \partial^2/\partial x^2 + \partial^2/\partial y^2 + \partial^2/\partial z^2$.

The stress boundary conditions for thermal stress problem may, in general, be given by $\sigma_{ij} n_j = 0$ on Γ, where the n_j are the direction cosines of the outward-drawn surface normal.

Next, we try to determine the solution of Eq. (4) by a method of successive approximations based upon the smallness of a fixed parameter ε defined [2] in the following form:

$$\mu_T = \mu_0\, e^{\varepsilon T} \tag{5}$$

where μ_0 and ε are material constants.

Using this parameter ε, the following expressions for the displacement are derived:

$$u_i = u_i^{(0)} + \varepsilon u_i^{(1)} + \varepsilon^2 u_i^{(2)} + \dots + \varepsilon^l u_i^{(l)} + \dots\,. \tag{6}$$

Substitution of Eq. (6) into Eq. (4) provides a set of successive boundary-value problems associated with terms of equal order in ε.

For the fundamental problem, which will be designated by the superscript 0, we find

$$\Delta u_i^{(0)} + \frac{1}{1-2\nu}\, e_{,i}^{(0)} = 2\,\frac{1+\nu}{1-2\nu} \left(\int_0^T \alpha_T(\eta)\, d\eta\right)_{,i}. \tag{7}$$

The boundary conditions may be given by $\sigma_{ij}^{(0)} n_j = 0$ on Γ. The displacement $u_i^{(0)}$ may be represented [4] by $u_i^{(0)} = \bar{u}_i^{(0)} + \bar{\bar{u}}_i^{(0)}$, where $\bar{u}_i^{(0)}$ is a particular solution, and $\bar{\bar{u}}_i^{(0)}$ is a complementary one. Here, we introduce the displacement potential $\Phi^{(0)}$ in the form

$$\bar{u}_i^{(0)} = \Phi_{,i}^{(0)}\,. \tag{8}$$

Substituting Eq. (8) into Eq. (7), we know that the system of Eq. (7) is satisfied if

$$\Delta \Phi^{(0)} = \frac{1+\nu}{1-\nu} \int_0^T \alpha_T(\eta)\, d\eta \tag{9}$$

and the corresponding stresses are

$$\bar{\sigma}_{ij}^{(0)} = 2\mu_T (\Phi_{,ij}^{(0)} - \delta_{ij}\, \Delta \Phi^{(0)})\,. \tag{10}$$

For a bounded body, it will in general be found that the solution $\Phi^{(0)}$ gives non-zero boundary tractions on the surface of the plate. The effects of removing those are found by solving an ordinary stress problem, using the homogeneous equations

$$\mu_T\, \Delta \bar{\bar{u}}_i^{(0)} + (\lambda_T + \mu_T)\, \bar{\bar{e}}_{,i}^{(0)} = 0\,. \tag{11}$$

By a similar argument we can construct the l-th boundary-value problem, which will be designated by the index l. We find

$$\Delta u_i^{(l)} + \frac{1}{1-2\nu}\, e_{,i}^{(l)} = -\frac{1}{\mu_T}\, T_{,j}\, \sigma_{ij}^{(l-1)} \tag{12}$$

and $\sigma_{ij}^{(l)} n_j = 0$ on Γ. The displacement $u_i^{(l)}$ may be expressed in terms of the four displacement functions $\Phi_i^{(l)}$ and $\Psi^{(l)}$ as follows:

$$u_i^{(l)} = 4(1-\nu)\, \Phi_i^{(l)} - \Psi_{,i}^{(l)} \tag{13}$$

where the functions $\Phi_i^{(l)}$ and $\Psi^{(l)}$ satisfy the equations

$$\Delta \Phi_i^{(l)} = -\frac{1}{4(1-\nu)}\, \frac{1}{\mu_T}\, T_{,j} \sigma_{ij}^{(l-1)}\,, \qquad \Delta \Psi^{(l)} = 2\Phi_{i,i}^{(l)} \tag{14}$$

and the corresponding stresses $\sigma_{ij}^{(l)}$ are

$$\sigma_{ij}^{(l)} = 2\mu_T[2\nu\delta_{ij}\Phi_{k,k}^{(l)} + 2(1-\nu)(\Phi_{j,i}^{(l)} + \Phi_{i,j}^{(l)}) - \Psi_{,ij}^{(l)}]\,. \tag{15}$$

Stress around a Crack in a Plate

Let a crack of length $2a$ be located in a plate under a linear temperature distribution in the X direction (Fig. 1) The plate is subjected to the temperature T_0 along $x = h$, and $-T_0$ along $x = -h$. We consider a state of plane strain.

If we postulate that the thermal conductivity K is temperature-dependent, the differential equation for the temperature field is given by div $(K \operatorname{grad} T) = 0$. In the two-dimensional case under consideration, we obtain $(KT_{,x})_{,x} + (KT_{,y})_{,y} = 0$ with the boundary conditions, $T(-h, y) = -T_0$ and $T(h, y) = T_0$. The mapping $\Theta = \left(\int_0^T K(\xi)\,d\xi\right)\Big/K_0$ yields a more convenient form of the foregoing equation, namely, $\Theta_{,xx} + \Theta_{,yy} = 0$ with the boundary conditions, $\Theta(-h, y) = -\Theta_0$ and $\Theta(h, y) = \Theta_0$, where $\Theta_0 = \left(\int_0^{T_0} K(\xi)\,d\xi\right)\Big/K_0$. The solution of the equation of heat conduction may be given by $\Theta = \Theta_0 x/h$. If we take the temperature dependence of the conductivity as $K = K_0(1 - kT)$ from [5], where k is a small constant, we find $T = [1 - (1 - 2k\Theta)^{1/2}]/k$. Considering the relation, $(1 - 2k\Theta)^{1/2} = 1 - (k\Theta_0 x/h) + O(k^2)$, the temperature field may be given approximately by $T = \Theta_0 x/h$. We approximate the variation of the coefficient of thermal expansion:

$$\alpha_T = \alpha_0(1 + \beta'_T T) \tag{16}$$

where α_0 are invariants.

Making a crack in the middle of the plate changes the stress distribution in the neighborhood of the crack. However, we can conclude from SAINT-VENANT's principle that the change is negligible at distances that are large compared with the crack length $2a$. Then, by calculating the resultant of the forces acting on a square plate, shown in Fig. 1 by the dashed lines, we can satisfy the following boundary conditions:

$$\begin{aligned}
&\int_{-h}^{h} \sigma_x\,dy = \int_{-h}^{h} y\sigma_x\,dy = \int_{-h}^{h} \tau_{xy}\,dy = 0 \quad \text{on} \quad x = \pm h\,;\\
&\int_{-h}^{h} \sigma_y\,dx = \int_{-h}^{h} x\sigma_y\,dx = \int_{-h}^{h} \tau_{xy}\,dx = 0 \quad \text{on} \quad y = \pm h\,.
\end{aligned} \tag{17}$$

It is obvious from the symmetry about the X axis that the problem of determining the stress distribution in the neighborhood of the crack is equivalent to that of determining the stress in the semiplate $y \geqq 0$ when the boundary $y = 0$ is subjected to the following conditions:

$$\tau_{xy} = \sigma_y = 0 \quad \text{for} \quad |x| < a\,, \qquad \tau_{xy} = u_y = 0 \quad \text{for} \quad |x| > a\,. \tag{18}$$

We now turn to the solution of the fundamental plane-strain problem, which can be obtained from Eqs. (9) and (11). Substituting Eq. (16) and the temperature distribution into Eq. (9), we obtain

$$\Phi^{(0)} = C_0(x^3/6 + \beta_T x^4/24) \tag{19}$$

where $C_0 = \{(1+\nu)/(1-\nu)\}\,\alpha_0(\Theta_0/h)$ and $\beta_T = \beta'_T(\Theta_0/h)$. The corresponding displacements and stresses are obtained from Eqs. (8) and (10) as

$$\bar{u}_x^{(0)} = C_0(x^2/2 + \beta_T x^3/6)\,, \qquad \bar{\sigma}_y^{(0)} = -2\mu_T C_0(x + \beta_T x^2/2)\,, \qquad \bar{u}_y^{(0)} = \bar{\sigma}_x^{(0)} = \bar{\tau}_{xy}^{(0)} = 0\,. \tag{20}$$

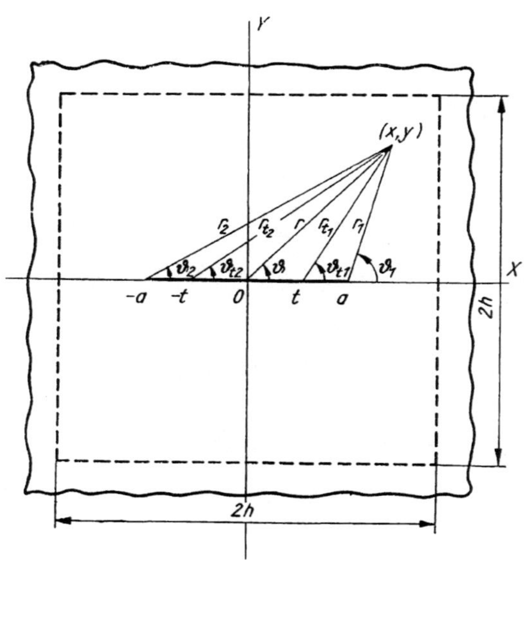

Fig. 1. Geometry of a finite crack in a plate.

Since the thermal stresses (20) do not satisfy the boundary conditions of Eqs. (17) and (18), we employ the solutions of Eq. (11):

$$\begin{aligned}
\bar{\bar{u}}_x^{(0)} &= -\nu C_1 x - \nu \frac{C_2 x^2}{2} - \frac{(\nu - 1)\, C_2 y^2}{z} + \frac{2}{\pi} \int_0^\infty (1 - 2\nu - \xi y)\, \{A(\xi) \cos(\xi x) - B(\xi) \sin(\xi x)\}\, e^{-\xi y}\, d\xi\,, \\
\bar{\bar{u}}_y^{(0)} &= (1 - \nu)\, C_1 y + (1 - \nu)\, C_2 x y + \frac{2}{\pi} \int_0^\infty (2 - 2\nu + \xi y)\, \{A(\xi) \sin(\xi x) + B(\xi) \cos(\xi x)\}\, e^{-\xi y}\, d\xi\,, \\
\bar{\bar{\sigma}}_x^{(0)} &= -2\mu_T \frac{2}{\pi} \int_0^\infty \xi (1 - \xi y)\, \{A(\xi) \sin(\xi x) + B(\xi) \cos(\xi x)\}\, e^{-\xi y}\, d\xi\,, \\
\bar{\bar{\sigma}}_y^{(0)} &= -2\mu_T \frac{2}{\pi} \int_0^\infty \xi (1 + \xi y)\, \{A(\xi) \sin(\xi x) + B(\xi) \cos(\xi x)\}\, e^{-\xi y}\, d\xi + 2\mu_T (C_1 + C_2 x)\,, \\
\bar{\bar{\tau}}_{xy}^{(0)} &= 2\mu_T \frac{2}{\pi}\, y \int_0^\infty \xi^2 \{A(\xi) \cos(\xi x) - B(\xi) \sin(\xi x)\}\, e^{-\xi y}\, d\xi\,.
\end{aligned} \tag{21}$$

The constants C_1 and C_2 may be determined from the boundary conditions as

$$C_1 = \frac{S_2^2 - S_1 S_3}{S_2 S_0 - S_1^2} \frac{\beta_T}{2} C_0\,, \qquad C_2 = \frac{\{S_1 (S_1 + \beta_T S_2/2) - S_0 (S_2 + \beta_T S_3/2)\}}{S_1^2 - S_0 S_2} C_0\,. \tag{22}$$

where $\beta_G = \beta'_G \Theta_0 / h$ and

$$\begin{aligned}
S_0 &= 2 \sinh(\beta_G h)/\beta_G\,, \qquad S_1 = 2h \cosh(\beta_G h)/\beta_G - 2 \sinh(\beta_G h)/\beta_G^2\,, \\
S_2 &= 2 \sinh(\beta_G h)\, (h^2 + 2/\beta_G^2)/\beta_G - 4h \cosh(\beta_G h)/\beta_G^2\,, \\
S_3 &= 2h \cosh(\beta_G h)\, (h^2 + 6/\beta_G^2)/\beta_G - 6 \sinh(\beta_G h)\, (h^2 + 2/\beta_G^2)/\beta_G^2\,.
\end{aligned} \tag{23}$$

It follows from Eqs. (21) that the boundary conditions (18) are satisfied if $A(\xi)$ and $B(\xi)$ satisfy the dual integral equations

$$\begin{aligned}
&\frac{2}{\pi} \int_0^\infty \xi A(\xi) \sin(\xi x)\, d\xi = -(C_0 - C_2)\, x && (|x| < a) \\
&\frac{2}{\pi} \int_0^\infty A(\xi) \sin(\xi x)\, d\xi = 0 && (|x| > a) \\
&\frac{2}{\pi} \int_0^\infty \xi B(\xi) \cos(\xi x)\, d\xi = C_1 - C_0 \frac{\beta_T}{2} x^2 && (|x| < a) \\
&\frac{2}{\pi} \int_0^\infty B(\xi) \cos(\xi x)\, d\xi = 0 && (|x| > a)\,.
\end{aligned} \tag{24}$$

The solution may be given [6] as

$$A(\xi) = \frac{\pi}{2} \frac{-C_0 + C_2}{2} \frac{a^2}{\xi} J_2(\xi a)\,, \qquad B(\xi) = \frac{\pi}{2} \left(C_1 - \frac{\beta_T a^2}{4} C_0\right) \frac{a}{\xi} J_1(\xi a) + \frac{\pi}{2} \frac{C_0 \beta_T}{4} \frac{2a^2}{\xi^2} J_2(\xi a)\,. \tag{25}$$

Substituting Eq. (25) into Eqs. (21) and writing $x + iy = r\, e^{i\theta}$, $(x - a) + iy = r_1\, e^{i\theta_1}$, and $(x + a) + iy = r_2\, e^{i\theta_2}$, in the notation of Fig. 1, we see that the expressions for the stress $\sigma_y^{(0)}$ may be put into the form

$$\begin{aligned}
\sigma_y^{(0)} = &-2\mu_T C_0 \left(x + \beta_T \frac{x^2}{2}\right) + 2\mu_T (C_1 + C_2 x) + \mu_T (C_0 - C_2)\, a^2 (I_1 + y I_2) - \\
&-2\mu_T \left(C_1 - \frac{\beta_T a^2}{4} C_0\right) a (I_3 + y I_4) - \mu_T C_0 \beta_T a^2 (I_5 + y I_6)
\end{aligned} \tag{26}$$

where

$$\begin{aligned}
I_1 &= \frac{1}{\sqrt{r_1 r_2}} \cos\left(\frac{\theta_1+\theta_2}{2}\right) + \frac{2r\cos\theta}{a^2} - \frac{2r^2}{\sqrt{r_1 r_2}}\,\frac{1}{a^2}\cos\left(2\theta - \frac{\theta_1+\theta_2}{2}\right),\\
I_2 &= \frac{3r}{\sqrt{(r_1 r_2)^3}} \sin\left\{\theta - \frac{3}{2}(\theta_1+\theta_2)\right\} - \frac{2r^3}{\sqrt{(r_1 r_2)^3}\,a^2} \sin\left\{3\theta - \frac{3}{2}(\theta_1+\theta_2)\right\},\\
I_3 &= \frac{1}{a}\left\{1 - \frac{r}{\sqrt{r_1 r_2}} \cos\left(\theta - \frac{\theta_1+\theta_2}{2}\right)\right\},\\
I_4 &= -\frac{a}{\sqrt{(r_1 r_2)^3}} \sin\left\{\frac{3}{2}(\theta_1+\theta_2)\right\},\\
I_5 &= \frac{1}{2} + \frac{r\sqrt{r_1 r_2}}{a^2} \cos\left(\theta + \frac{\theta_1+\theta_2}{2}\right) - \frac{r^2}{a^2}\cos 2\theta\,,\\
I_6 &= \frac{1}{\sqrt{r_1 r_2}} \sin\left(\frac{\theta_1+\theta_2}{2}\right) - \frac{2r}{a^2}\sin\theta + \frac{2r^2}{\sqrt{r_1 r_2}\,a^2} \sin\left(2\theta - \frac{\theta_1+\theta_2}{2}\right).
\end{aligned} \tag{27}$$

The stress intensity factors of the fundamental problem at $x = a$ and $x = -a$ are given by

$$\begin{aligned}
K_I^{(0)}|_{x=a} &= \lim_{x\to a+} \sqrt{2}\,\sqrt{x-a}\;\sigma_y^{(0)} = 2\sqrt{a}\,k_1\,e^{\beta_G a}\,\{C_1 + (-C_0 + C_2)\,a/2 - C_0\beta_T a^2/4\}\,,\\
K_I^{(0)}|_{x=-a} &= \lim_{x\to -a-} \sqrt{2}\,\sqrt{-x-a}\;\sigma_y^{(0)} = 2\sqrt{a}\,k_1\,e^{-\beta_G a}\,\{C_1 - (-C_0 + C_2)\,a/2 - C_0\beta_T a^2/4\}\,.
\end{aligned} \tag{28}$$

If the constant $\beta_T = 0$, that is, the coefficient of thermal expansion α_T is not a function of temperature, Eq. (28) will yield a zero value for $K_I^{(0)}$. But $\beta_G = 0$ will yield a nonzero value of $K_I^{(0)}$. Thus, whether $K_I^{(0)}$ is zero or not depends only on the temperature variation of the function α_T.

By a similar argument we can construct the first boundary-value problem in plane strain from Eqs. (14). We can find the solution of Eqs. (14), and the corresponding displacements and stresses may be obtained from Eqs. (13) and (15) as

$$\begin{aligned}
u_x^{(1)} &= \frac{1}{1-\nu}\,\frac{\Theta_0}{h}\,\frac{1}{4\pi}\int_0^\infty e^{-\xi y}\,\{4(1-\nu)\,(\xi y^2 - y) - y + (-3+2\nu)\,\xi^{-1}\}\,\{A(\xi)\sin\xi x + B(\xi)\cos\xi x\}\,d\xi\, -\\
&\quad - (1-\nu)\,C_3 x + \frac{2}{\pi}\int_0^\infty (-1 + 2\nu + \xi y)\,e^{-\xi y}\,\{B_1(\xi)\sin\xi x - B_2(\xi)\cos\xi x\}\,d\xi\,,\\
u_y^{(1)} &= \frac{1}{1-\nu}\,\frac{\Theta_0}{h}\,\frac{1}{4\pi}\int_0^\infty e^{-\xi y}\,\{4(1-\nu)\,(\xi y^2 + y) - y + (-2+2\nu)\,\xi^{-1}\}\,\{A(\xi)\cos\xi x - B(\xi)\sin\xi x\}\,d\xi\, +\\
&\quad + \nu C_3 y + \frac{2}{\pi}\int_0^\infty (2 - 2\nu + \xi y)\,e^{-\xi y}\,\{B_1(\xi)\cos\xi x + B_2(\xi)\sin\xi x\}\,d\xi\,,\\
\sigma_x^{(1)} &= \frac{2\mu_T}{1-\nu}\,\frac{\Theta_0}{h}\,\frac{1}{4\pi}\int_0^\infty e^{-\xi y}\,\{4(1-\nu)\,(\xi^2 y^2 - \xi y) - \xi y + 4\nu - 3\}\,\{A(\xi)\cos\xi x - B(\xi)\sin\xi x\}\,d\xi\, -\\
&\quad - 2\mu_T C_3 - 2\mu_T\,\frac{2}{\pi}\int_0^\infty \xi(1 - \xi y)\,e^{-\xi y}\,\{B_1(\xi)\cos\xi x + B_2(\xi)\sin\xi x\}\,d\xi\,,\\
\sigma_y^{(1)} &= \frac{2\mu_T}{1-\nu}\,\frac{\Theta_0}{h}\,\frac{1}{4\pi}\int_0^\infty e^{-\xi y}\,\{4(1-\nu)\,(-\xi^2 y^2 + \xi y) + \xi y + 5 - 4\nu\}\,\{A(\xi)\cos\xi x - B(\xi)\sin\xi x\}\,d\xi\, -\\
&\quad - 2\mu_T\,\frac{2}{\pi}\int_0^\infty \xi(1 + \xi y)\,e^{-\xi y}\,\{B_1(\xi)\cos\xi x + B_2(\xi)\sin\xi x\}\,d\xi\,,\\
\tau_{xy}^{(1)} &= \frac{2\mu_T}{1-\nu}\,\frac{\Theta_0}{h}\,\frac{1}{4\pi}\int_0^\infty e^{-\xi y}\,\xi y\,\{4(1-\nu)\,(1 - \xi y) + 1\}\,\{A(\xi)\sin\xi x + B(\xi)\cos\xi x\}\,d\xi\, -\\
&\quad - 2\mu_T\,\frac{2}{\pi}\,y\int_0^\infty e^{-\xi y}\,\xi^2\,\{B_1(\xi)\sin\xi x - B_2(\xi)\cos\xi x\}\,d\xi
\end{aligned} \tag{29}$$

where $A(\xi)$ and $B(\xi)$ are given by Eqs. (25).

The functions $B_1(\xi)$ and $B_2(\xi)$ and the constant C_3 may be determined from the boundary conditions. The boundary conditions are $\tau_{xy}^{(1)} = \sigma_y^{(1)} = 0$ for $|x| < a$ and $\tau_{xy}^{(1)} = u_y^{(1)} = 0$ for $|x| > a$; and $\int_{-h}^{h} \sigma_x^{(1)}\,dy = \int_{-h}^{h} y\sigma_x^{(1)}\,dy = \int_{-h}^{h} \tau_{xy}^{(1)}\,dy = 0$ on $x = \pm h$ and $\int_{-h}^{h} \sigma_y^{(1)}\,dx = \int_{-h}^{h} x\sigma_y^{(1)}\,dx = \int_{-h}^{h} \tau_{xy}^{(1)}\,dx = 0$ on $y = \pm h$. The constant C_3 in Eqs. (29) is given by

$$C_3 = \frac{1}{1-\nu}\frac{\Theta_0}{h^2}\frac{1}{4\pi}\int_0^h dy \int_0^\infty d\xi\, e^{-\xi y}\,\{4(1-\nu)(\xi^2 y^2 - \xi y) - \xi y + 4\nu - 3\}\,\{A(\xi)\cos\xi x - B(\xi)\sin\xi x\}\,. \tag{30}$$

The functions $B_1(\xi)$ and $B_2(\xi)$ in Eqs. (29) must satisfy the dual integral equations

$$\begin{aligned}
&\int_0^\infty \xi D_1(\xi)\cos\xi x\,d\xi = \frac{\Theta_0}{2h}\int_0^\infty A(\xi)\cos\xi x\,d\xi \qquad (|x| < a)\\
&\int_0^\infty (2-2\nu)\,D_1(\xi)\cos\xi x\,d\xi = 0 \qquad (|x| > a)\\
&\int_0^\infty \xi D_2(\xi)\sin\xi x\,d\xi = -\frac{\Theta_0}{2h}\int_0^\infty B(\xi)\sin\xi x\,d\xi \qquad (|x| < a)\\
&\int_0^\infty (2-2\nu)\,D_2(\xi)\sin\xi x\,d\xi = 0 \qquad (|x| > a)
\end{aligned} \tag{31}$$

where

$$D_1(\xi) = B_1(\xi) - \frac{1}{1-\nu}\frac{\Theta_0}{8h}\frac{A(\xi)}{\xi}, \qquad D_2(\xi) = B_2(\xi) + \frac{1}{1-\nu}\frac{\Theta_0}{8h}\frac{B(\xi)}{\xi}. \tag{32}$$

Eqs. (31) completely determine the functions $B_1(\xi)$ and $B_2(\xi)$ as shown in Eqs. (24). We find

$$\begin{aligned}
D_1(\xi) &= \frac{\pi}{2}\frac{\Theta_0}{4h}(-C_0 + C_2)\,a^2\xi^{-2}J_2(\xi a)\,,\\
D_2(\xi) &= -\frac{\pi}{2}\frac{\beta_T\Theta_0}{16h}\,a^2\xi^{-1}\,[\{4(C_1 + C_0\xi)^{-2} - C_0a^2\}\,J_2(\xi a) - C_0a\xi^{-1}J_1(\xi a)]\,.
\end{aligned} \tag{33}$$

Substituting Eqs. (33) into (29), we can determine the displacements and the corresponding stresses. The stress intensity factor of the first boundary-value problem is given by

$$K_I^{(1)}|_{x=a} = e^{\beta_G a}\frac{\Theta_0}{h}\frac{a\sqrt{a}}{8}\{-4C_1 + \beta_T C_0 a^2/2\}\,, \qquad K_I^{(1)}|_{x=-a} = e^{-\beta_G a}\frac{\Theta_0}{h}\frac{a\sqrt{a}}{8}\{-4C_1 + \beta_T C_0 a^2/2\}\,. \tag{34}$$

Then, the total stress intensity factor is given from Eqs. (28) and (34) by

$$K_I = K_I^{(0)} + \varepsilon K_I^{(1)}\,. \tag{35}$$

Similarly, we can solve the l-th boundary value problem.

Numerical Results and Discussion

To show the convergence of the solution numerically and clarify the effects of the temperature-dependent properties, we performed the numerical calculations for a plate with $h/a = 10$, made of steel with following properties [2]: $\beta_G = -0.008652$, $\beta_T = 0.02919$, $\nu = 0.35$. The computations were carried out with the aid of a FACOM 230-75 electronic computer at Nagoya University. The numerical integrations were performed using SIMPSON's formula.

Fig. 2 shows the variation of σ_y with x and y. In this figure, the stress σ_y is equal to the sum of $\sigma_y^{(0)}$ and $\sigma_y^{(1)}$; the dashed curves show the stress distribution of $\sigma_y^{(1)}$. From the results, it may be seen that the solution of the fundamental problem represents a major contribution to the complete solution of the problem, while the contribution of the first boundary-value problem is almost insignificant.

The maximum stress of σ_y occurs at the tip of the crack. It is seen that the maximum stress is rapidly decreasing as the y/a is increasing and the magnitude of the maximum stress at $x = -a$ is larger than that of the maximum stress at $x = a$. For example, when $y/a = 0.2$ the maximum stress at $x = a$ is 3 per cent less than the maximum stress at $x = -a$. This effect may be caused by the temperature dependency of the shear modulus.

Fig. 3 shows the isochromatic lines $\tau/[\{(1+\nu)/(1-\nu)\}\,\alpha_0\Theta_0\mu_0] =$ const. in the vicinity of the crack for the temperature-sensitive stress field, where $\tau = |(\sigma_y - \sigma_x)/2 + i\tau_{xy}|$. On account of the temperature dependence of the material properties, the stress distribution around the tip of the crack $x = a$ is different from the distribution around $x = -a$. This difference may be mainly caused by the temperature dependence of the shear modulus.

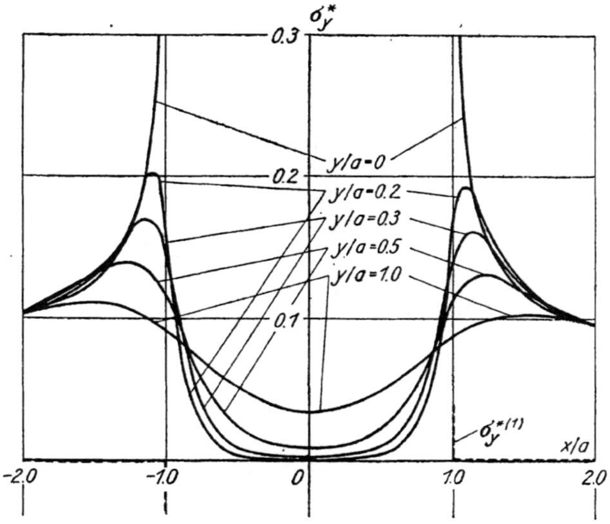

Fig. 2. Variation of σ_y^* with x and y
[$\sigma_y^* = \sigma_y/\{(1+\nu)/(1-\nu)\,\alpha_0\Theta_0\mu_0\}$]

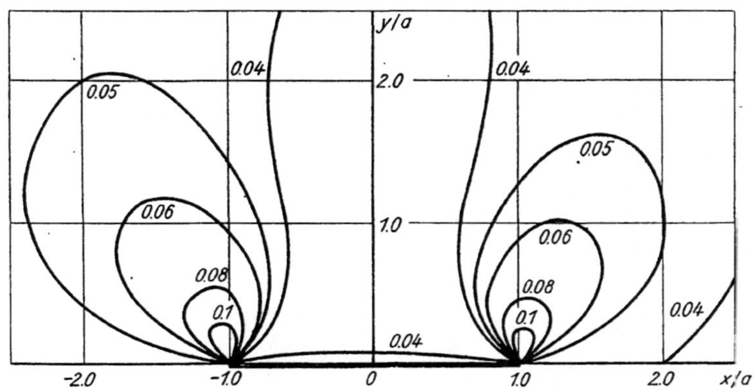

Fig. 3. Isochromatic lines of $\tau/\{(1+\nu)/(1-\nu)\,\alpha_0\Theta_0\mu_0\}$ = Const.

Finally, we obtained the stress intensity factor $K_I^* = K_I\{(1+\nu)\,\alpha_0\Theta_0\mu_0/(1-\nu)\}$ as follows:

$K_I^*|_{x=a} = 0.9479\,E-1\,,\qquad K_I^*|_{x=-a} = 0.9670\,E-1$ for the sensitive plate,

$K_I^*|_{x=a} = K_I^*|_{x=-a} = 0$ for the insensitive plate.

It should be noted that the stress intensity factor depends on the temperature-dependent properties of the plate.

Acknowledgement

The author is especially grateful to Prof. Dr. A. ATSUMI, Tohoku University and Prof. Dr. Y. TAKEUTI, University of Osaka prefecture, for his invaluable direction.

References

1 TROSTEL, R., Stationäre Wärmespannungen mit temperaturabhängigen Stoffwerten, Ing. Arch. **26** (1958), 416–434.
2 NYUKO, H., TAKEUTI, Y., NODA, N., Thermal Stresses in a Composite Hollow Circular Cylinder with Temperature Dependency of Material Properties, Jour. Japan Soc. Mech. Engrs. **44** (1978), 1454–1460 (in Japanese).
3 HATA, T., Thermoelastic Problem for a GRIFFITH Crack in a Plate with Temperature-Dependent Properties under a Linear Temperature Distribution, J. Thermal Stresses **3** (1979), 353–366.
4 TAKEUTI, Y., Analyses of Thermal Stresses, Nissin, Tokyo 1971 (in Japanese).
5 NOWINSKI, J., Transient Thermoelastic Problem for an Infinite Medium with a Spherical Cavity Exhibiting Temperature-Dependent Properties, J. Appl. Mech. **29** (1962), 399–407.
6 SNEDDON, I. N., LOWENGRUB, M., Crack Problems in the Classical Theory of Elasticity, Wiley, New York, 1969.

Eingereicht am 15. 3. 1980

Anschrift: Dr. TOSHIAKI HATA, Faculty of Education, Shizuoka University, Shizuoka City, 422, Japan

ZAMM 61, 89 – 97 (1981)

N. PHAN-THIEN / C. J. GOH

On the Fibre Pull-out Problem

In this paper, the force required to pull out a long, slender fibre embedded in a semi-infinite elastic matrix by a distance W is derived. Assuming that the fibre is slender and that there is no slip at the fibre-matrix interface, it is shown that the nature of the inclusion depends on the parameter $\frac{E^*}{E}\left(\frac{R}{l}\right)^2 \ln\left(\frac{2l}{R}\right)$, *where E, E* are the* YOUNG's *moduli of the matrix and the fibre respectively. R is a typical radius of the fibre and l its length. When this parameter is large, the fibre is effectively rigid and the load transfer occurs over the entire length of the fibre. If this parameter is very much less than unity, then the inclusion is elastic in nature and the load transfer occurs in a finite neighbourhood of the embedded end of the fibre which is near the free surface of the semi-infinite matrix.*

В данной работе определяется сила, необходимая для вырывания длинного тонкого волокна, конечная часть которого длиной W уложена в полубесконечном упругом связующем материале. В предположении, что волокно тонкое и что поверхность соприкосновения волокна со связующим материалом свободна от скольжения, показывается, что по своей природе такое включение зависит от параметра $\frac{E^*}{E}\left(\frac{R}{l}\right)^2 \ln\left(\frac{2l}{R}\right)$, где E и E^* — модули Юнга связующего материала и волокна соответственно. R — типичный радиус волокна и l — его длина. Если этот параметр велик, то волокно ведёт себя эффективно как жесткое и сила переноса распределяется по всей длине волокна. Если этот параметр намного меньше единицы, включение ведёт себя как упругое и сила переноса оказывается в конечном соседстве с уложенным концом волокна, который расположен вблизи свободной поверхности полубесконечного связующего материала.

In dieser Arbeit wird die erforderliche Kraft bestimmt, die notwendig ist, um eine lange, dünne Fiber, die in einem halbunendlichen elastischen Bindemittel über der Länge W eingebettet ist, auszureißen. Unter der Annahme, daß die Fiber dünn ist und daß kein Schlupf auf der Fiber-Bindemittel-Grenzschicht auftritt, wird gezeigt, daß die typische Natur der Einschließung von dem Parameter $\frac{E^*}{E}\left(\frac{R}{l}\right)^2 \ln \frac{2l}{R}$ *abhängt, wo E und E* die Elastizitätsmodule des Bindemittels und der Fiber bezeichnen. R ist ein typischer Radius der Fiber und l ihre Länge. Ist dieser Parameter groß, so verhält sich die Fiber effektiv als starr, und die Übertragungskraft verteilt sich über die gesamte Länge der Fiber. Wenn dieser Parameter sehr viel kleiner als 1 ist, verhält sich die Einschließung elastisch, und die Übertragungskraft erscheint in endlicher Nachbarschaft vom eingebetteten Ende der Fiber, wobei diese nahe an der freien Oberfläche des halbunendlichen Bindemittels liegt.*

1. Introduction

In this communication, we consider the problem of a fibre being pulled out from a semi-infinite elastic matrix (Fig. 1). Earliest analyses of this problem appear to be given by MUKI and STERNBERG [1, 2, 3] who solved certain FREDHOLM integral equations numerically to obtain the axial load transfer characteristics. A semi-analytical solution was given by LAWRENCE [4] who assumed a shear stress distribution along the fibre-matrix interface. The incentive to study this simple problem becomes obvious when referring to Figure 5.21 of KELLY [5] which illustrates the crack initiation phase of a fibre-reinforced material. During this phase, unbroken fibres serve as links between the two crack faces; these will be pulled out from the matrix during the crack propagation phase. Thus, the understanding of the single fibre pull-out problem will, hopefully, provide some insight into this complicated phenomenon.

Analytical progress is possible if the fibre aspect ratio (l/R) is large so that the slender-body theory developed by BATCHELOR [6] and other authors [7—10] is applicable. This asymptotic theory has been applied by RUSSEL [11], RUSSEL and ACRIVOS [12], PHAN-THIEN [13—15] and PHAN-THIEN and HUILGOL [16] to solve a number of problems of fibrereinforced solids. In essence, this theory involves replacing the actual fibre by a line distribution of concentrated forces and force doublets (KELVIN state [17]). With a proper choice of the line density of the KELVIN-state distribution, the no-slip condition at the fibrematrix interface can approximately be satisfied. In particular, the rigid fibre pull-out problem has been solved [13] by replacing the embedded part of the fibre with a line distribution of MINDLIN-state.

The layout of the paper is as follows: First, MINDLIN's solution of a concentrated force near a free boundary and the slender-body theory will be discussed. Then, the analysis of the pull-out problem will be given for two cases: the "rigid" and the "elastic" fibre.

2. Slender-Body Theory — Mindlin's Solution

A good starting point for our analysis is the reciprocal theorem introduced by LORENTZ [18]. This theorem has been presented by BRENNER [19] and RAMKISSOON [20]. The solid analogy is reproduced here without proof.

At any point in the solution domain, the NAVIER-CAUCHY equation requires that, neglecting inertial forces,

$$\sigma_{ij,j} + f_i = 0\,, \tag{2.1}$$

where $\boldsymbol{f}$ is the body force per unit volume, $\boldsymbol{\sigma}$ is the stress tensor given by

$$\sigma_{ij} = \lambda u_{k,k}\delta_{ij} + \eta(u_{i,j} + u_{j,i})\,. \tag{2.2}$$

Here λ, η are the LAMÉ constants of the solid, $\boldsymbol{u}$ is the displacement vector. Cartesian tensor notation with the comma denoting covariant derivative is used.

A solution to (2.1—2.2) satisfying some relevant boundary conditions is denoted by $\{\boldsymbol{u}, \boldsymbol{\sigma}, \boldsymbol{f}\}$.

The reciprocal theorem asserts that, if $\{\boldsymbol{u}, \boldsymbol{\sigma}, \boldsymbol{f}\}$ and $\{\boldsymbol{u}', \boldsymbol{\sigma}', \boldsymbol{f}'\}$ are two possible solutions to (2.1—2.2) then

$$\int_S (\sigma_{ij}u_i' - \sigma_{ij}'u_i)\, n_j \,\mathrm{d}S = \int_V (f_i'u_i - f_iu_i')\,\mathrm{d}V\,, \tag{2.3}$$

where V is a volume in the solution domain and S is its bounding surface.

The basic idea behind slender-body theory is that the actual fibre is replaced by a suitably chosen distribution of MINDLIN-states [1]. By MINDLIN-state, we mean the effects of a concentrated force near the presence of a free boundary*). Mathematically, for the "dash" solution we choose

$$f_i'(\boldsymbol{x}) = \delta_{ik}\delta(\boldsymbol{x} - \boldsymbol{y})\,,$$

i.e., a unit concentrated force at $\boldsymbol{y}$ in the k direction. We denote the "dash" displacement and stress by $u_{ik}^{(M)}$ and $\sigma_{ijk}^{(M)}$ respectively, where the superscript M denotes MINDLIN-state solution [21].

When the body force is neglected, the reciprocal theorem (2.3) gives

$$u_k(\boldsymbol{x}) = \int_S (\sigma_{ij}u_{ik}^{(M)} - \sigma_{ijk}^{(M)}u_i)\, n_j \,\mathrm{d}s\,. \tag{2.4}$$

Equation (2.4) is similar to BETTI's theorem [22] in the theory of elasticity. The only difference between (2.4) and BETTI's theorem is that KELVIN's solution has now been replaced by MINDLIN's solution to take into account the effects of the free boundary.

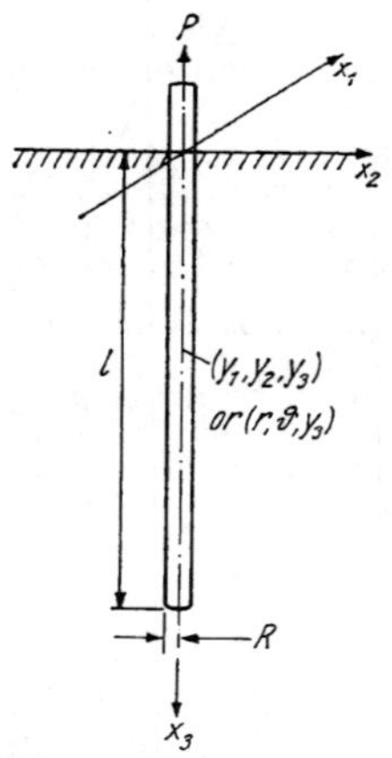

Fig. 1. Geometry of the Fibre Pull-out Problem

In (2.4) the integral is carried over the surface of the embedded fibre (Fig. 1). Since the fibre is slender, on its surface we have

$$y_1, y_2 \sim O(R); \qquad y_3 \sim O(l)\,.$$

Hence we can carry out a TAYLOR expansion on the MINDLIN displacement and stress tensor to obtain

$$u_{ik}^{(M)}(\boldsymbol{y}) = u_{ik}^{(M)}\Big|_0 + \frac{\partial u_{ik}^{(M)}}{\partial y_1}\Big|_0 y_1 + \frac{\partial u_{ik}^{(M)}}{\partial y_2}\Big|_0 y_2 + O(y_1^2, y_2^2) \tag{2.5}$$

and

$$\sigma_{ijk}^{(M)}(\boldsymbol{y}) = \sigma_{ijk}^{(M)}\Big|_0 + \frac{\partial \sigma_{ijk}^{(M)}}{\partial y_1}\Big|_0 y_1 + \frac{\partial \sigma_{ijk}^{(M)}}{\partial y_2}\Big|_0 y_2 + O(y_1^2, y_2^2)\,, \tag{2.6}$$

where the subscript 0 denotes an evaluation at $y_1 = 0 = y_2$.

If we further require that the variation in y_3 is much less than variations in y_1 and y_2 then the displacement outside the fibre is given by

$$u_k(x) = \int_0^l \left\{ f_i u_{ik}^{(M)}\Big|_0 + (g_{i1}\delta_{1l} + g_{i2}\delta_{2l} - h_{il})\frac{\partial u_{ik}^{(M)}}{\partial y_l}\Big| + \ldots \right\} \mathrm{d}y_3\,, \tag{2.7}$$

where

$$f_i(y_3) = \int_0^{2\pi} \sigma_{ij}n_j\, R\,\mathrm{d}\theta\,, \qquad g_{il}(y_3) = \int_0^{2\pi} \sigma_{ij}y_l n_j R\,\mathrm{d}\theta\,,$$

$$h_{il}(y_3) = \int_0^{2\pi} [\lambda u_n n_n \delta_{il} + \eta(u_i n_l + u_l n_i)]\, R\,\mathrm{d}\theta\,, \qquad R = (y_1^2 + y_2^2)^{1/2}\,.$$

*) We call this singularity MINDLIN-state because the original solution of a concentrated force near the presence of a free boundary was discovered by MINDLIN [21] and that the analogous singularity in an infinite medium is called KELVIN-state by GURTIN [17].

As pointed out by Russel and Acrivos [12], $\boldsymbol{f}$ is the line density of the Mindlin-state (force per unit length exerted on the fibre) while $\boldsymbol{g}$ and $\boldsymbol{h}$ are the densities of the force doublets. Physically, the fibre has now been replaced by a line distribution of forces and force doublets.

Our present objective is to evaluate the unknown functions $\boldsymbol{f}$, $\boldsymbol{g}$, $\boldsymbol{h}$, which seems to be a formidable task, but a close examination of the geometry of the problem, as suggested by Russel [11], reveals that the only non-zero functions are f_3, $g_{11} = g_{22} = g$, and h_{33}. Furthermore, h_{33} can be absorbed into f_3 [11] and thus we are left with two unknown scalar functions g and $f = h'_{33} + f_3$ and (2.7) becomes

$$u_k(\boldsymbol{x}) = \int\limits_0^l \left\{ u_{3k}^{(M)}\Big|_0 f + g\left(\frac{\partial u_{1k}^{(M)}}{\partial y_1}\Big|_0 + \frac{\partial u_{2k}^{(M)}}{\partial y_2}\Big|_0\right)\right\} \mathrm{d}y_3 . \tag{2.8}$$

Mindlin's solution is tabulated in Appendix A for the convenience of the reader. Detailed discussion of this solution is provided in the original paper by Mindlin [21] or in a monograph by Westergaard [24].

In cylindrical co-ordinates the displacement field outside the fibre becomes

$$u_r = \frac{r(1+\nu)}{8\pi E(1-\nu)} \{ I_3^a(f) + (3-4\nu)\, J_3^a(f) + 4(1-\nu)\,(1-2\nu)\,\xi_{11}(f) - 6xJ_3^{b1}(f) - 4\nu I_3(g) + $$
$$+ (12\nu - 8)\, J_3(g) + 24xJ_5^1(g) + 16(1-\nu)\,(1-2\nu)\,\xi_{12}(g)\} , \tag{2.9}$$

and

$$u_3 = \frac{1+\nu}{8\pi E(1-\nu)} \{ (3-4\nu)\, I_1(f) + [8(1-\nu)^2 - (3-4\nu)]\, J_1(f) + I_3^{a^2}(f) + (3-4\nu)\, J_3^{b^2}(f) - 2xJ_3^1(f) + $$
$$+ 6xJ_5^{b_1^2}(f) - 2I_3^a(g) - 2(3-4\nu)\, J_3^a(g) + 8(1-\nu)\,(1-2\nu)\,\xi_{11}(g) - 12xJ_5^{b1}(g)\} + O(r^2) , \tag{2.10}$$

where, from now on, unless otherwise stated, we conveniently write x for x_3, y for y_3 and r for $(x_1^2 + x_2^2)^{1/2}$.

The integrals in expressions (2.9) and (2.10) are defined as follows

$$I_n(f) = \int\limits_0^l \frac{f(y)\,\mathrm{d}y}{(r^2 + (x-y)^2)^{n/2}}, \qquad J_n(f) = \int\limits_0^l \frac{f(y)\,\mathrm{d}y}{(r^2 + (x+y)^2)^{n/2}}, \qquad n = 1, 3, 5$$

$$I_m^{a^n}(f) = \int\limits_0^l \frac{(x-y)^n\, f(y)\,\mathrm{d}y}{(r^2 + (x-y)^2)^{m/2}}, \qquad J_m^{a^n}(f) = \int\limits_0^l \frac{(x-y)^n\, f(y)\,\mathrm{d}y}{(r^2 + (x-y)^2)^{m/2}}, \qquad n = 1, 2; \qquad m = 1, 3, 5$$

$$J_m^n(f) = \int\limits_0^l \frac{y^n f(y)\,\mathrm{d}y}{(r^2 + (x+y)^2)^{m/2}}, \qquad n = 1, 3; \quad m = 1, 3, 5$$

$$J_k^{b^n m}(f) = \int\limits_0^l \frac{(x+y)^n\, y^m f(y)\,\mathrm{d}y}{(r^2 + (x+y)^2)^{k/2}}, \qquad n = 1, 2; \quad m = 0, 2; \quad k = 1, 3, 5$$

$$\xi_{11}(f) = \int\limits_0^l \frac{f(y)\,\mathrm{d}y}{(r^2 + (x+y)^2)^{1/2}\,(x + y + ((x+y)^2 + r^2)^{1/2})},$$

$$\xi_{12}(f) = \int\limits_0^l \frac{f(y)\,\mathrm{d}y}{(r^2 + (x+y)^2)^{1/2}\,(x + y + (r^2 + (x+y)^2)^{1/2})^2}.$$

There are no exact solutions of (2.9)—(2.10) available. However, the slenderness of the fibre enables us to carry out asymptotic analysis on these integrals with the slenderness ratio R/l as the appropriate perturbation parameter. Similar asymptotic analysis has been given by Tillet [7] and Fraenkel [24] and is given in Appendix B for the interested reader.

Once the displacement field in the matrix is determined, the corresponding stress field can be calculated:

$$\sigma_{rr} = \lambda\Theta + 2\eta\frac{\partial u_r}{\partial r}, \qquad \sigma_{r3} = \sigma_{3r} = \eta\left(\frac{\partial u_r}{\partial x} + \frac{\partial u_3}{\partial r}\right), \qquad \sigma_{\theta\theta} = \lambda\Theta + 2\eta\frac{u_r}{r}, \qquad \sigma_{33} = \lambda\Theta + 2\eta\frac{\partial u_3}{\partial x} \tag{2.11}$$

where the Lamé constants λ and η are related to the Young modulus, E, and Poisson's ratio, ν, of the matrix by

$$\lambda = \frac{E\nu}{(1+\nu)\,(1-2\nu)}; \qquad \eta = \frac{E}{2(1+\nu)}.$$

In (2.11), Θ is the divergence of the displacement field:

$$\Theta = \nabla\cdot\boldsymbol{u} = \frac{\partial u_r}{\partial r} + \frac{u_r}{r} + \frac{\partial u_3}{\partial x}.$$

The traction on the fibre-matrix interface is thus given by

$$T_i = \sigma_{ij} n_j \quad \text{where} \quad \boldsymbol{n} = (1, 0, -\mathrm{d}R/\mathrm{d}x) + O(R^2)\,,$$

so that, if we only consider cylindrical fibres, then

$$T_r = \sigma_{rr} n_r = \sigma_{rr}\,, \qquad T_3 = \sigma_{r3} n_r = \sigma_{r3}\,. \tag{2.12}$$

The force and force doublet densities (f, g) can be found by requiring that the normal tractions T_r and T_3 and the displacements u_3 and u_r are continuous across the fibre-matrix interface. Within the fibre, the displacements are given by (RUSSEL [11]):

$$u_3 = C(x)\,, \qquad u_r = rB(x) \tag{2.13}$$

from which the traction inside the fibre can be determined as

$$T_r = \lambda^* \left(2B + \frac{\mathrm{d}C}{\mathrm{d}x}\right) + 2\eta^* B + O(E^* R^2)\,, \qquad T_3 = \eta^* r \frac{\mathrm{d}B}{\mathrm{d}x} + O(E^* R^2) \tag{2.14}$$

where λ^* and η^* are the LAMÉ constants of the fibre material.

3. "Nearly-Rigid" Fibre

The analysis of a rigid fibre pulled out from a semi-infinite matrix has previously be done by PHAN-THIEN [13] where the function f was found to be a constant in the leading term of the asymptotic analysis and g is neglected. In this section, we extend the domain of validity of [1] by showing that the rigid fibre solution is correct as long as the parameter

$$\frac{E^*}{E}\left(\frac{R}{l}\right)^2 \ln\left(\frac{2l}{R}\right)$$

is very much geater than unity. This fact has been pointed out by RUSSEL and ACRIVOS [12] in a slightly different context.

The asymptotic analyses of various integrals tabulated in Section 2 are given in Appendix B using the matched-asymptotic expansion proposed by FRAENKEL [24] and TILLET [7]. It should be noted at this stage that the asymptotic results do not hold at the ends of the fibre ($x = 0, l$) where stress singularities may be present.

The asymptotic displacement field is given by

$$\begin{aligned} u_r(\boldsymbol{x}) = \frac{r(1+\nu)}{8\pi E(1-\nu)} \Bigg\{ & -2f' \ln \frac{2\sqrt{x(l-x)}}{r} + (3-4\nu) \frac{fl}{(x+l)^2} + 2(1-\nu)(1-2\nu) \frac{fl}{x(x+l)} - \\ & - \frac{xf(3x-l)\,l^2}{x^2(x+l)^3} - 4\nu\left(\frac{2g}{r^2} + \frac{1}{2} g'' \ln \frac{2\sqrt{x(l-x)}}{r}\right) + (12\nu - 8) \frac{gl(2x+l)}{x^2(x+l)^2} + \\ & + 24g\left[\frac{(x+l)^3 - x^3}{3x^2(x+l)^3} - \frac{(x+l)^4 - x^4}{4x^2(x+l)^4}\right] + 2(1-\nu)(1-2\nu) \frac{gl(2x+l)}{x^2(x+l)^2} \Bigg\} \end{aligned} \tag{3.1}$$

and

$$\begin{aligned} u_3(\boldsymbol{x}) = \frac{1+\nu}{8\pi E(1-\nu)} \Bigg\{ & 8(1-\nu)\, f \ln \frac{2\sqrt{x(l-x)}}{r} + 8g' \ln \frac{2\sqrt{x(l-x)}}{r} + \\ & + 2\left(\frac{2(1-\nu)(1-2\nu)\,l}{x(x+l)} - \frac{(3x-l)\,l^2}{x(x+l)^3} - \frac{(3-4\nu)\,l}{(x+l)^2}\right) g \Bigg\}. \end{aligned} \tag{3.2}$$

Away from the ends of the fibre ($x = 0, l$) and retaining only highest order terms, (3.1) and (3.2) reduce to

$$u_r(\boldsymbol{x}) = \frac{r(1+\nu)}{8\pi E(1-\nu)} \left\{ -2f' \ln \frac{2\sqrt{x(l-x)}}{r} + \frac{(4\nu^2 - 6\nu + 3)\, f}{x} - \frac{8\nu g}{r^2} \right\}, \tag{3.3}$$

$$u_3(\boldsymbol{x}) = \frac{1+\nu}{8\pi E(1-\nu)} \left\{ 8(1-\nu)\, f \ln \frac{2\sqrt{x(l-x)}}{r} + 4g' \ln \frac{2\sqrt{x(l-x)}}{r} + \frac{2(4\nu^2 - 6\nu + 3)\, g}{x} \right\}. \tag{3.4}$$

After differentiating the displacement field and substituting the results into (2.12), the stress and hence the tractions become:

$$T_3 = \sigma_{r3} = \frac{-1}{8\pi(1-\nu)} \left\{ \frac{4(1-\nu)\, f}{r} + \frac{2(1+2\nu)\, g'}{r} \right\}, \tag{3.5}$$

$$\begin{aligned} T_r = \sigma_{rr} = {} & \frac{\nu}{8\pi(1-\nu)(1-2\nu)} \left\{ 4(1-2\nu)\, f' \ln \frac{2\sqrt{x(l-x)}}{r} + \frac{2f(4\nu^2 - 8\nu + 5)}{x} + \frac{2(4\nu^2 - 6\nu + 4)}{x} \right\} + \\ & + \frac{1}{8\pi(1-\nu)} \left\{ -2f' \ln \frac{2\sqrt{x(l-x)}}{r} + \frac{(4\nu^2 - 6\nu + 3)\, f}{x} + \frac{8\nu g}{r^2} \right\}. \end{aligned} \tag{3.6}$$

These results can be matched with (2.13) and (2.14) to determine the nature of the inclusion and the order of magnitude of the f and g and their derivatives. Matching of u_r and u_3 gives

$$C(x) = \frac{1+\nu}{8\pi E(1-\nu)} \left\{ 8(1-\nu) f \ln \frac{2\sqrt{x(l-x)}}{R} + 4g' \ln \frac{2\sqrt{x(l-x)}}{R} + \frac{2g(4\nu^2 - 6\nu + 3)}{x} \right\}, \tag{3.7}$$

$$B(x) = \frac{1+\nu}{8\pi E(1-\nu)} \left\{ -2f' \ln \frac{2\sqrt{x(l-x)}}{R} + \frac{4\nu^2 - 6\nu + 3}{x} f - \frac{8\nu g}{R^2} \right\}. \tag{3.8}$$

Upon substitution into (2.14) the tractions on the fibre become

$$\begin{aligned} T_r = {} & \frac{E^*\nu^*(1+\nu)}{(1+\nu^*)(1-2\nu^*)\, 8\pi E(1-\nu)} \left\{ -4f' \ln \frac{2\sqrt{x(l-x)}}{R} + \frac{2(4\nu^2-6\nu+3) f}{x} - \frac{16\nu g}{R^2} + \frac{4(1-\nu) f}{x} + \right. \\ & \left. + 8(1-\nu) f' \ln \frac{2\sqrt{x(l-x)}}{R} + 4g'' \ln \frac{2\sqrt{x(l-x)}}{R} + \frac{2g'}{x} - \frac{2g(4\nu^2-6\nu+3)}{x^2} + \frac{2g'(4\nu^2-6\nu+3)}{x} \right\} + \\ & + \frac{E^*(1+\nu)}{(1+\nu^*)\, 8\pi E(1-\nu)} \left\{ -2f' \ln \frac{2\sqrt{x(l-x)}}{R} + \frac{(4\nu^2-6\nu+3) f}{x} - \frac{8\nu g}{R^2} \right\} \end{aligned} \tag{3.9}$$

and

$$T_3 = \frac{-E^*(1+\nu)}{16(1+\nu^*)\pi E(1-\nu)} \left\{ 2f'' R \ln \frac{2\sqrt{x(l-x)}}{R} + \frac{f' R}{x} + \frac{(4\nu^2-6\nu+3) f R}{x^2} - \frac{(4\nu^2-6\nu+3) f' R}{x} + \frac{8\nu g'}{R} \right\}. \tag{3.10}$$

Taking only highest order terms and noticing that f and g are relatively smooth functions with little fluctuation, we have, for the matching of T_r

$$\begin{aligned} & \frac{E^*\nu^*(1+\nu)}{8\pi E(1+\nu^*)(1-2\nu^*)(1-\nu)} \left\{ -4f' \ln \frac{2\sqrt{x(l-x)}}{R} + 2\frac{4\nu^2-6\nu+3}{x} f - \frac{16\nu g}{R^2} + \frac{4(1-\nu) f}{x} + \right. \\ & \left. + 8(1-\nu) f' \ln \frac{2\sqrt{x(l-x)}}{R} \right\} + \frac{E^*(1+\nu)}{8\pi E(1+\nu^*)(1-\nu)} \left\{ -2f' \ln \frac{2\sqrt{x(l-x)}}{R} + \frac{4\nu^2-6\nu+3}{x} f - \frac{8\nu g}{R^2} \right\} \sim \\ & \sim \frac{\nu}{8\pi(1-\nu)(1-2\nu)} \left\{ 4(1-2\nu) f' \ln \frac{2\sqrt{x(l-x)}}{R} + 2\frac{4\nu^2-8\nu+5}{x} f + 2\frac{4\nu^2-6\nu+4}{x} g' \right\} + \\ & + \frac{1}{8\pi(1-\nu)} \left\{ -2f' \ln \frac{2\sqrt{x(l-x)}}{R} + \frac{4\nu^2-6\nu+3}{x} f + \frac{8\nu g}{R^2} \right\}, \end{aligned} \tag{3.11}$$

while the matching of T_3 gives

$$\begin{aligned} & \frac{E^*(1+\nu)}{16\pi E(1+\nu^*)(1-\nu)} \left\{ \frac{f' R}{x} + \frac{4\nu^2-6\nu+3}{x^2} f R - \frac{4\nu^2-6\nu+3}{x} f' R - \frac{8\nu g'}{R} \right\} \sim \\ & \sim \frac{1}{8\pi(1-\nu)} \left\{ \frac{4(1-\nu) f}{R} + \frac{2(1+2\nu) g'}{R} \right\}. \end{aligned} \tag{3.12}$$

Thus, when

$$\frac{E^*}{E} \left(\frac{R}{l}\right)^2 \ln\left(\frac{2l}{R}\right) \gg 1,$$

or when the fibre is nearly rigid, (3.11) gives

$$f' = 0\left(\frac{g}{R^2 \ln(2l/R)}\right), \tag{3.13}$$

whereas (3.12) gives

$$g' = O(R^2 f), \quad g = O(R^2 f/x) \tag{3.14}$$

so that we would expect f to be nearly constant away from the ends of the fibre.

In view of (3.13) and (3.14) the leading term in (3.4) is

$$u_3(x) = \frac{1+\nu}{\pi E} f \ln \frac{2\sqrt{x(l-x)}}{R} \sim \frac{1+\nu}{\pi E} f \ln \frac{2l}{R}. \tag{3.15}$$

In deriving the asymptotic result of (3.15) we have neglected a term

$$\ln\left(\frac{x}{l}\left(1 - \frac{x}{l}\right)\right)^{1/2}$$

which takes a value of approximately -1 away from the fibre ends.

Since f remains constant away from the ends, u_3 is also constant and if the relative displacement between the fibre and the matrix is W, then f can be expressed as

$$f = \frac{\pi E W}{(1+\nu)\ln\left(\frac{2l}{R}\right)}. \tag{3.16}$$

The error in this approximation is

$$O\left(\left(\ln\frac{2l}{R}\right)^{-1}\right).$$

Consequently the pull-out force is given by

$$P = \int_0^l f\,\mathrm{d}x = \frac{\pi E W l}{(1+\nu)\ln\left(\frac{2l}{R}\right)} \tag{3.17}$$

which is identical to the expression given by PHAN-THIEN [13] who assumed a rigid fibre.

This expression shows that the load transfer between the matrix and fibre occurs over the entire embedded length of the fibre. The interfacial shear stress away from the fibre ends also remains constant and is given by

$$\sigma_{r3} = -\frac{f}{2\pi R} = -\frac{EW}{2(1+\nu)\,R\ln(2l/R)} \tag{3.18}$$

as shown in [13] where the rigid fibre is replaced by a line distribution of forces.

Note also that our asymptotic analysis fails at the ends of the fibre. However, in the case of a near-rigid fibre, the load transfer occurs over the entire length of the fibre; consequently, integral properties (e.g., pull-out force) do not depend on the end conditions. Hence, the errors involved in (3.17) and (3.18) are of the order $O((\ln 2l/R)^{-1})$.

4. Elastic Fibre Case

The problem of an elastic fibre completely embedded in an infinite matrix has been investigated by RUSSEL [11]. It was found that the asymptotic limits for the traction are not strictly valid at the ends, but in the elastic limit, i.e., when

$$\frac{E^*}{E}\left(\frac{R}{l^2}\right)^2 \ln\frac{2l}{R} \ll 1\,,$$

the results indicate that for a fibre of uniform radius throughout its length, $f = 0$ except at the ends where axial load transfer occurs. The solution provided by RUSSEL [11] can be verified in a slightly different manner by assuming that

$$g = g_0R^2 = \text{constant}\,, \qquad f = R^2[f_0\delta(x) + f_1\delta(l-x)]\,. \tag{4.1), (4.2}$$

Assuming this, we can evaluate the integrals in (2.9) and (2.10) for two cases: when $x \ll R$, i.e. near the pull-out end, and $x \gg R$, i.e. away from the pull-out end. The calculation is straightforward but tedious and is shown in Appendix C.

Substituting the results of Appendix C into (2.9)–(2.10) and retaining only highest order terms, the radial and axial displacement fields are reduced to, when $x \ll R$,

$$u_r(x) = \frac{(1+\nu)\,R^2}{8\pi E(1-\nu)\,r}\{4(1-\nu)\,(1-2\nu)\,f_0 + 16(1-\nu)^2\,g_0\} \tag{4.3}$$

and

$$u_3(x) = \frac{R^2(1-\nu^2)}{\pi E r}\{f_0 + g_0\}\,. \tag{4.4}$$

Away from the free surface, $x \gg R$, the resulting displacement fields are

$$u_r(x) = \frac{r(1+\nu)\,R^2}{8\pi E(1-\nu)}\left\{\frac{f_0}{x^2} + \frac{f_1}{(x-l)^2} + (3-4\nu)\left[\frac{f_0}{x^2} + \frac{(x-l)\,f_1}{(x+l)^3}\right] + 2(1-\nu)\,(1-2\nu)\left[\frac{f_0}{x^2} + \frac{f_1}{(x+l)^2}\right] - \right.$$
$$\left. - \frac{6xlf_1}{(x+l)^4} - \frac{8\nu g_0}{r^2} + 8g_0\left[\frac{(x+l)^3 - x^3}{x^2(x+l)^3}\right] + 2(1-\nu)\,(1-2\nu)\,\frac{l(2x+l)}{x^2(x+l)^2}\right\} \tag{4.5}$$

and

$$u_3(x) = \frac{(1+\nu)\,R^2}{8\pi E(1-\nu)}\left\{4(1-\nu)\,(1-2\nu)\,\frac{f_0}{x} + 4(1-\nu)\,\frac{f_1}{x-l} + 8(1-\nu)^2\,\frac{f_1}{x+l} + \frac{4xlf_1}{(x+l)^3}\right\} -$$
$$- \frac{2(1+\nu)\,R^2l}{8\pi E(1-\nu)}\left\{\frac{1}{x(x-l)} + \frac{(3-4\nu)\,(2x+l)}{x(x+l)^2} - \frac{2(1-\nu)\,(1-2\nu)}{x(x+l)} + \frac{l(3x-l)}{x(x+l)^3}\right\} g_0\,. \tag{4.6}$$

If the fibre is long, we expect no load transfer at the embedded end of the fibre, i.e. $f_1 = 0$. Furthermore, in the region $R \ll x \ll l$, (4.5) and (4.6) can be further reduced, up to the leading terms, to

$$u_r(x) = \frac{r(1+\nu)\,R^2}{4\pi E(1-\nu)}\left\{\frac{(3-2\nu)\,(1-\nu)\,f_0}{x^2} - \frac{4\nu g_0}{r^2}\right\}, \tag{4.7}$$

$$u_3(x) = \frac{R^2}{x}\left\{\frac{(1+\nu)\,(3-2\nu)\,f_0}{2\pi E} + \frac{(1+\nu)\,(4\nu^2-2\nu+1)\,g_0}{(1-\nu)\,4\pi E}\right\}. \tag{4.8}$$

Consequently, with the use of (1.15) and (1.17), the tractions can be easily shown to be

$$T_3 = \sigma_{3r} = -\frac{(3-2\nu)\,R^2 f_0 r}{4\pi x^3}, \tag{4.9}$$

$$T_r = \sigma_{rr} = \frac{\nu R^2 g_0}{\pi(1-\nu)\,r^2} + \frac{R^2 f_0(3-2\nu)}{4\pi x^2}. \tag{4.10}$$

Matching of (2.13)—(2.14) and (4.7)—(4.10) gives

$$u_3 = C(x) = \frac{R^2}{x}\left\{\frac{(1+\nu)\,(3-2\nu)\,f_0}{2\pi E} + \frac{(1+\nu)\,(4\nu^2-2\nu+1)\,g_0}{(1-\nu)\,4\pi E}\right\}, \tag{4.11}$$

$$u_r = B(x) = \frac{(1+\nu)\,R^2}{4\pi E(1-\nu)}\left\{\frac{(3-2\nu)\,(1-\nu)\,f_0}{x^2} - \frac{8\nu g_0}{R^2}\right\}. \tag{4.12}$$

A substitution of (4.11) and (4.12) into (2.14) gives, for the matching of T_3 and T_r,

$$T_r = g_0\left\{\frac{\nu}{\pi(1-\nu)\,R} + \frac{E^*\nu^*}{(1+\nu^*)\,(1-2\nu^*)}\left[\frac{(1+\nu)\,\nu}{2\pi E(1-\nu)\,R^2} + \frac{(1+\nu)\,(4\nu^2-2\nu+1)}{(1-\nu)\,4\pi E}\right] - \right.$$
$$\left. - \frac{E^*(1+\nu)\,\nu}{(1+\nu^*)\,\pi E(1-\nu)\,R^2}\right\} \sim \left[\frac{E^*(1+\nu)}{E(1+\nu^*)} + 1\right]\frac{(3-2\nu)\,f_0}{4\pi x^2}, \tag{4.13}$$

$$T_3 = -\frac{E^*(1+\nu)}{E(1+\nu^*)}\,\frac{(3-2\nu)\,f_0 R^3}{4\pi x^3} \sim -\frac{(3-2\nu)\,f_0 R^3}{4\pi x^3}. \tag{4.14}$$

(4.13)—(4.14) indicate that

$$g_0 \sim O(R^2 f_0) \tag{4.15}$$

and

$$\frac{E^*(1+\nu)}{E(1+\nu^*)} = \frac{\eta^*}{\eta} \sim O(1) \tag{4.16}$$

from which it can be seen that

$$\frac{E^*}{E}\left(\frac{R}{l}\right)^2 \ln\frac{2l}{R} \ll 1\,. \tag{4.17}$$

Furthermore, since (4.15) indicates that g_0 is much smaller than f_0, the displacement of the fibre very near to the free boundary is

$$u_3 = \frac{R(1-\nu^2)\,f_0}{\pi E}. \tag{4.18}$$

Hence, if we pull out the fibre by a distance W, small enough to avoid debonding, then

$$f_0 = \frac{W\pi E}{R(1-\nu^2)}, \tag{4.19}$$

and consequently,

$$f(x) = \frac{\pi E W R\,\delta(x)}{1-\nu^2}, \tag{4.20}$$

and thus the pull-out force P and the shear stress are given by

$$P = \int_0^l f(y)\,\mathrm{d}y = \frac{\pi E W R}{1-\nu^2}, \tag{4.21}$$

and

$$\sigma_{r3} = -\frac{(3-2\nu)\,W E R^2}{4(1-\nu^2)\,x^3}. \tag{4.22}$$

Note that the pull-out force is independent of the fibre length and closely resembles the solution of the rigid cylindrical punch problem, where it was found, through a solution of an appropriate integral equation (see LITTLE

[25]), that the force required to effect an impression W by the punch into the elastic matrix is given by

$$P = \frac{2REW}{1-\nu^2} \tag{4.23}$$

where R is the radius of the cylindrical punch.

5. Concluding Remarks

In summary, we have asymptotically solved the elastic fibre pull-out problem. The nature of the inclusion depends on the parameter $\frac{E^*}{E}\left(\frac{R}{l}\right)^2 \ln\left(\frac{2l}{R}\right)$. If this parameter is much greater than unity then we have essentially a rigid fibre and the load transfer takes place over the entire fibre length. In this case the interfacial shear stress remains constant over the fibre length except at the fibre ends where stress singularities may exist. In this regard, HANDELSMAN and KELLER [26], TILLET [7] and RUSSEL and ACRIVOS [12] have asserted that the fibre ends should not be blunter than a prolate spheroid to assure the validity of the asymptotic analysis up to the order $O(R/l)$. We expect this to be the case here also. On the other hand if $\frac{E^*}{E}\left(\frac{R}{l}\right)^2 \ln\left(\frac{2l}{R}\right) \ll 1$, then the inclusion is elastic in nature and the load transfer occurs only in a finite neighbourhood of the fibre end which is near the free surface. In this case, the shear stress rapidly decays to zero as $1/x^3$, where x is the distance from the free surface.

In practice, a slight variation in the radius of the fibre and its cross-sectional geometry should be expected. This has been discussed in detail by BATCHELOR [6], TUCK [10], RUSSEL [11] and RUSSEL and ACRIVOS [12]. Although we have not discussed an arbitrarily shaped fibre in this present work, its inclusion presents no particular problem. Specifically, a slight variation in the fibre radius should not affect the leading term in the asymptotic expansion of a "nearly-rigid" fibre, and the load transfer in the "elastic" fibre can be shown to be proportional to $\mathrm{d}R^2/\mathrm{d}x$ as pointed out by RUSSEL [11].

This present analysis may be useful in estimating the so-called elastic "pull-out fracture toughness" which is needed in understanding fracture mechanisms in fibrous materials.

Acknowledgements

We thank Dr. W. B. RUSSEL of Princeton University for his helpful suggestions. NP-T wishes to acknowledge an IRAC grant from the Internal Research Assessment Committee of the University of Newcastle. This work forms part of an honour BE-thesis by C. J. GOH which has been submitted to the University of Newcastle, Australia.

Appendix A: Mindlin's Solution

By superimposing various GALERKIN vectors, MINDLIN [21] solved the three-dimensional elasticity equations for the case of a unit concentrated force acting in the interior of a semi-infinite solid. The results are summarised as follows:

$$U_{3k}^{(M)}(\boldsymbol{x}) = \frac{(x_k - y_k)(1+\nu)}{8\pi E(1-\nu)}\left\{\frac{(x_3-y_3)}{R_1^3} + \frac{(3-4\nu)(x_3-y_3)}{R_2^3} + \frac{4(1-\nu)(1-2\nu)}{R_2(R_2+x_3+y_3)} - \frac{6y_3x_3(x_3+y_3)}{R_2^5}\right\}, \qquad k = 1, 2$$

$$U_{kk}^{(M)}(\boldsymbol{x}) = \frac{(1+\nu)}{8\pi E(1-\nu)}\left\{\frac{(3-4\nu)}{R_1} + \frac{1}{R_2} + \frac{(x_k-y_k)^2}{R_1^3} + \frac{(3-4\nu)(x_k-y_k)^2}{R_2^3} + \frac{2y_3x_3}{R_2^3}\left(1 - \frac{3(x_k-y_k)^2}{R_2^2}\right) + \right.$$
$$\left. + \frac{4(1-\nu)(1-2\nu)}{(R_2+x_3+y_3)}\left(1 - \frac{(x_k-y_k)^2}{R_2(R_2+x_3+y_3)}\right)\right\}, \qquad k = 1, 2$$

$$U_{33}^{(M)}(\boldsymbol{x}) = \frac{1+\nu}{8\pi E(1-\nu)}\left\{\frac{(3-4\nu)}{R_1} + \frac{8(1-\nu)^2-(3-4\nu)}{R_2} + \frac{(x_3-y_3)^2}{R_1^3} + \frac{(3-4\nu)(x_3+y_3)^2 - 2y_3x_3}{R_2^3} + \frac{6x_3y_3(x_3+y_3)^2}{R_2^5}\right\}$$

$$U_{12}^{(M)}(\boldsymbol{x}) = U_{21}^{(M)}(\boldsymbol{x}) = \frac{(x_1-y_1)(x_2-y_2)(1+\nu)}{8\pi E(1-\nu)}\left\{\frac{1}{R_1^3} + \frac{3-4\nu}{R_2^3} - \frac{6y_3x_3}{R_2^5} - \frac{4(1-\nu)(1-2\nu)}{R_2(R_2+x_3+y_3)^2}\right\},$$

$$U_{k3}^{(M)}(\boldsymbol{x}) = \frac{(x_k-y_k)(1+\nu)}{8\pi E(1-\nu)}\left\{\frac{(x_3-y_3)}{R_1^3} + \frac{(3-4\nu)(x_3-y_3)}{R_2^3} - \frac{4(1-\nu)(1-2\nu)}{R_2(R_2+x_3+y_3)} + \frac{6x_3y_3(x_3+y_3)}{R_2^5}\right\}, \qquad k = 1, 2$$

where

$$R_1^2 = (x_1-y_1)^2 + (x_2-y_2)^2 + (x_3-y_3)^2, \qquad R_2^2 = (x_1-y_1)^2 + (x_2-y_2)^2 + (x_3+y_3)^2,$$

and $U_{ij}^M(\boldsymbol{x})$ denotes the displacement at position vector $\boldsymbol{x}$ in the i direction due to a concentrated force acting in the j direction at position vector $\boldsymbol{y}$.

Appendix B: Asymptotic Results of Integrals for "Rigid" Fibre

The integrals appearing in Eqn. (2.9) and (2.10) can be divided into two categories — those that involve the term $[(x-y)^2+r^2]^{1/2}$ and those that involve $[(x+y)^2+r^2]^{1/2}$ in the denominator of the integrands. The former case represents a singular integral when $x = y$ and is commonly encountered in previous works particularly in the field of slender body Fluid Mechanics, e.g. TUCK [9], TILLET [7] or HENDELSMAN and KELLER [26]. Basically, the methods of matched asymptotic expansion due to FRAENKEL [24] where inner and outer expansion operators are used to expand the integrals with the slenderness ratio as the perturbation parameter.

In the latter case, a straightforward TAYLOR's expansion is carried out in the integrand and the result is asymptotically evaluated. This procedure is also partially justified when we found out later that f and g are approximately constant away from the ends. For illustration purposes, we will show one simple case of $J_1(f)$.

By definition,

$$J_1(f) = \int_0^l \frac{f(y)\,dy}{(r^2 + (x+y)^2)^{1/2}} = \int_{-l}^0 \frac{f(-y)\,dy}{(r^2 + (x-y)^2)^{1/2}}.$$

However, f is still undefined on the negative y axis so we can arbitrarily define f to be an even function, so that if we let

$$J_1(f) + I_1(f) = I(f)$$

where

$$I_1(f) = \int_0^l \frac{f(y)\,dy}{(r^2 + (x-y)^2)^{1/2}} = 2f(x) \ln \frac{2\sqrt{x(l-x)}}{r}$$

and (TILLET [7])

$$I(f) = \int_{-l}^l \frac{f(y)\,dy}{(r^2 + (x-y)^2)^{1/2}} = 2f(x) \ln \frac{2\sqrt{l^2 - x^2}}{r}$$

then

$$J_1(f) = I(f) - I_1(f) = f(x) \ln \left(1 + \frac{l}{x}\right)$$

which is identical to letting $f(y) = f(x) + \dots$ and asymptotically evaluating the integral

$$J_1(f) = \int_0^l \frac{f(x)}{x+y} \left[1 + \left(\frac{r}{x+y}\right)^2\right]^{-1/2} dy \cong f(x) \ln \left(1 + \frac{l}{x}\right) + O(r^2).$$

This argument can be applied similarly to the rest of the latter integrals. The calculations involved are not difficult but tedious, only the summary is given below (retaining only highest order terms for each):

$$I_3^a(f) = 2f(x) \ln \frac{2\sqrt{x(l-x)}}{r}, \qquad J_3^a(f) = \frac{f(x)\,l}{(x+l)^2}, \qquad \xi_{11}(f) = \frac{f(x)\,l}{2x(x+l)},$$

$$J_5^{b1}(f) = \frac{(3x-l)\,l^2 f(x)}{6x^2(x+l)^3}, \qquad I_3(g) = \frac{2g(x)}{r^2} + \frac{1}{2} g''(x) \ln \frac{2\sqrt{x(l-x)}}{r}, \qquad J_3(g) = \frac{g(x)\,l(2x+l)}{x^2(x+l)^2},$$

$$J_5^1(g) = \frac{(x+l)^3 - x^3}{3x^3(x+l)^3} - \frac{(x+l)^4 - x^4}{4x^3(x+l)^4}, \qquad \xi_{12}(g) = \frac{g(x)\,l(2x+l)}{8x^2(x+l)^2}, \qquad I_3^{a2}(f) = 2f(x) \ln \frac{2\sqrt{x(l-x)}}{r},$$

$$J_3^{b^2}(f) = f(x) \ln \left(1 + \frac{l}{x}\right), \qquad J_3^1(f) = \frac{l^2 f(x)}{2x(x+l)^2}, \qquad J_5^{b_1^2}(f) = \frac{l^2 f(x)}{2x(x+l)^2},$$

$$I_3^a(g) = -2g'(x) \ln \frac{2\sqrt{x(l-x)}}{r}, \qquad J_3^a(g) = \frac{g(x)\,l}{(x+l)^2}, \qquad \xi_{11}(g) = \frac{g(x)\,l}{2x(x+l)}, \qquad J_5^{b1}(g) = \frac{g(x)\,(3x-l)\,l^2}{6x^2(x+l)^3}.$$

Appendix C: Asymptotic Results of Integrals for "Elastic" Fibre

Evaluation of the integrals in this case is straightforward with the assumptions of (4.1) and (4.2). In most cases exact close form solutions are available but we only require the highest order terms. So again we asymptotically expand the close form solution and retain only the leading terms. Two ranges of the fibres are considered: very near to the end, $x \ll R$ and away from the ends, $R \ll \ll x \ll l$. Again, only asymptotic results are tabulated as follows:

	"very close to the ends"	"away from the ends"
$I_3^a(f)$	$R^2\left[\frac{xf_0}{r^3} - \frac{f_1}{l^2}\right]$	$R^2\left[\frac{f_0}{x^2} + \frac{f_1}{(x-l)^2}\right]$
$J_3^a(f)$	$R^2\left[\frac{f_0 x}{r^3} - \frac{f_1}{l^2}\right]$	$R^2\left[\frac{f_0}{x^2} + \frac{(x-l)\,f_1}{(x+l)^3}\right]$
$\xi_{11}(f)$	$R^2\left[\frac{f_0}{r^2}\right]$	$R^2\left[\frac{f_0}{2x^2} + \frac{f_1}{2(x+l)^2}\right]$
$J_5^{b1}(f)$	$R^2\left[\frac{f_1}{l^3}\right]$	$R^2\left[\frac{f_1 l}{(4+l)^4}\right]$
$I_3(g)$	$R^2\left[\frac{g_0}{r^2}\right]$	$R^2\left[\frac{2g_0}{r^2}\right]$
$J_3(g)$	$R^2\left[\frac{g_0}{r^2}\right]$	$R^2\left[\frac{g_0(2xl + l^2)}{2x^2(x+l)^2}\right]$
$J_5^1(g)$	$R^2\left[\frac{g_0}{3r^2}\right]$	$R^2 g_0\left[\frac{1}{3}\left(\frac{1}{x^3} - \frac{1}{(x+l)^3}\right) + \frac{x}{4}\left[\frac{1}{x^4} - \frac{1}{(x+l)^4}\right]\right]$
$\xi_{12}(g)$	$R^2\left[\frac{g_0}{2r^2}\right]$	$R^2\left[\frac{g_0 l(2x+l)}{8x^2(x+l)^2}\right]$
$I_1(f)$	$R^2\left[\frac{f_0}{r^2}\right]$	$R^2\left[\frac{f_0}{x} + \frac{f_1}{x-l}\right]$

$J_1(f)$	$R^2\left[\frac{f_0}{r^2}\right]$	$R^2 \frac{f_0}{x} + \left[\frac{f_1}{x+l}\right]$
$I_3^{a2}(f)$	$R^2\left[\frac{f_1}{l}\right]$	$R^2\left[\frac{f_0}{x} + \frac{f_1}{x-l}\right]$
$J_3^{b2}(f)$	$R^2\left[\frac{f_1}{l}\right]$	$R^2\left[\frac{f_0}{x} + \frac{f_1}{x+l}\right]$
$J_3^1(f)$	$R^2\left[\frac{f_1}{l^2}\right]$	$R^2\left[\frac{f_1 l}{(x+l)^3}\right]$
$J_5^{b2}(f)$	$R^2\left[\frac{f_1}{l^2}\right]$	$R^2\left[\frac{f_1 l}{(x+l)^3}\right]$
$I_3^a(g)$	$R^2\left[-\frac{g_0}{r}\right]$	$R^2\left[\frac{g_0 l}{x(x-l)}\right]$
$J_3^a(g)$	$R^2\left[-\frac{g_0}{r}\right]$	$R^2\left[\frac{g_0 l(2x+l)}{x(x+l)^2}\right]$
$\xi_{11}(g)$	$R^2\left[-\frac{g_0}{r}\right]$	$R^2\left[\frac{g_0 l}{2x(x+l)}\right]$
$J_5^{b1}(g)$	$R^2\left[\frac{g_0}{3r^2}\right]$	$R^2\left[\frac{g_0(3x-l)\,l^2}{6x^2(x+l)^3}\right]$

where the error for the case of "very close to the ends" is of the order $0(R^2x^2)$ and $0(R^4)$ for the case of "away from the ends".

References

1 Sternberg, E. / Muki, R., Load absorption by a filament in a fibre reinforced material, Z. Ang. Math. und Physik **21**, 552 (1970).
2 Sternberg, E. / Muki, R., On the diffusion of an axial load from an infinite cylindrical bar embedded in an elastic medium, Int. J. Solids Structure **5**, 587 (1969).
3 Sternberg, E. / Muki, R., Elastoplastic load-transfer to half space from a partially-embedded axially loaded rod, Int. J. Solids Structure **6**, 69 (1970).
4 Lawrence, P., Some theoretical consideration of fibre pull-out from an elastic matrix, J. Material Sci. **7**, 1 (1972).
5 Kelly, K., Strong Solids, Clarendon Press, Oxford (1973).
6 Batchelor, G. K., Slender-body theory for particles of arbitrary cross-section in Stokes flow, J. Fluid Mech. **44**, 529 (1970).
7 Tillet, J. P. K., Axial and transverse Stokes flow past slender axisymmetric bodies, J. Fluid Mech. **44**, 401 (1970).
8 Cox, R. G., The motion of long slender bodies in a viscous fluid, I and II, J. Fluid Mech. **44**, 791 (1970), **45**, 625 (1971).
9 Tuck, E. O., Some methods for flows past blunt slender bodies, J. Fluid Mech. **18**, 619 (1964).
10 Tuck, E. O., Toward the calculation and minimization of Stokes drag on bodies of arbitrary shape, Proc. 3rd Australasian Conf. on Hydraulics and Fluid Mechanics, November (1968).
11 Russel, W. B., On the effective moduli of composite materials: Effect of fibre length and geometry at dilute concentrations, Z. Ang. Math. und Physik **24**, 581 (1973).
12 Russel, W. B. and Acrivos, A., On the effective moduli of composite materials, slender rigid inclusion at dilute concentration, Z. Ang. Math. und Physik **23**, 434 (1972).
13 Phan-Thien, N., A contribution to the rigid fibre pull-out problem, Fibre Sci. and Tech., **3** 179 (1980).
14 Phan-Thien, N., The inclusion problem for a rigid slender body in an infinite elastic medium, Fibre Sci. and Tech. **12**, 235 (1979).
15 Phan-Thien, N., A constitutive equation for a dilute suspension of nearly straight, flexible fibres in an elastic medium, Fibre Sci. and Tech., **13**, 303 (1980).
16 Phan-Thien, N. / Huilgol, R., A micromechanic theory of chopped-fibre-reinforced materials, Fibre Sci. and Tech., **13** 423 (1980).
17 Gurtin, M. E., Handbuch der Physik, Band V 1a/2 (Ed. C. Truesdell), Springer-Verlag (1972).
18 Lorentz, H. A., Amsterdam, Zittingsverlag Akad. v. Wet. **5**, 168 (1896), see also Abhandlungen über Theor. Physik, p. 23, Teubner (1906).
19 Brenner, H., The Stokes resistance of an arbitrary particle, I, Chem. Eng. Sci. **18**, 1 (1963).
20 Ramkissoon, H., Representation in hydrodynamics, Acta Mechanica **28**, 49 (1971).
21 Mindlin, R. D., Force at a point in the interior of a semi-infinite solid, Physics **7**, 195 (1936).
22 Kupradze, V. D., Potential Methods in the Theory of Elasticity, Daniel Davey and Co. (1965).
23 Westergaard, H. M., Theory of Elasticity and Plasticity, Harvard University (1972).
24 Fraenkel, L. E., On the methods of matched asymptotic expansion, Parts I, II and III, Proc. Cambridge Phil. Soc. **65**, 209 (1965).
25 Little, R. W., Elasticity, Prentice Hall (1973).
26 Handelsman, R. A. / Keller, J. B., Axially symmetric potential flow around a slender body, J. Fluid Mech. **28**, 131 (1967).

Eingereicht am 17. 9. 1979, in revidierter Fassung am 2. 6. 1980

Anschrift: Dr. N. Phan-Thien, University of Sydney, Department of Mechanical Engineering, Sydney, New South Wales 2006, Australia

ZAMM 61, 99 – 104 (1981)

S. K. DEY

Error Propagations in Implicit Discretized Viscous Flow Models

Im allgemeinen wird in Anspruch genommen, daß implizite endliche Differenzenlösungen der Navier-Stokesschen Gleichungen unbedingt stabil sind. Wie eine linearisierte Stabilitätsanalyse [1] zeigte, ist dieser Anspruch gerechtfertigt, wenn die endlichen Differenzendarstellungen der Randbedingungen keine Fehler enthalten und wenn die Abbruchfehler vernachlässigt werden. In den meisten Problemen werden jedoch die Randbedingungen nur näherungsweise erfaßt und somit entstehen hier Fehler. Weiterhin können die Abbruchfehler in Abhängigkeit von den Fließbedingungen und Eingangsparametern (wie z. B. Schrittweite in der Zeit, Gitterform) einen signifikanten Einfluß auf die numerischen Lösungen haben. In der vorliegenden Arbeit werden hierzu entsprechende Untersuchungen durchgeführt.

It has been claimed in general, that implicit finite difference solution of Navier-Stokes' equations is unconditionally stable. A linearized stability analysis [1] showed that if finite difference representations of boundary conditions do not contain errors and if truncation errors are neglected, such a claim is justifiable. However, in most problems, boundary conditions are often approximated and hence errors originate from there. Furthermore, depending upon the flow conditions and input parameters (like time steps, mesh sizes), truncation errors may have significant contributions upon numerical solutions. In this paper these investigations have been made.

В общем случае принято предполагать, что неявные конечные разностные решения уравнений Навье-Стокса являются необходимо устойчивыми. Как показывает линеаризированный анализ устойчивости [1], это предположение оправдано, если конечные разностные представления граничных условий не содержат ошибок и если можно пренебречь погрешностью от обрыва. Во многих задачах граничные условия учитываются, однако, только приближённым образом, в результате чего и возникают ошибки. Кроме того, большое влияние на численные решения могут иметь погрешности от обрыва в зависимости от условий течения и от параметров на входе (например, ширина шага во времени, форма решётки). В настоящей работе проведены соответствующие исследования.

1. Introduction

Analysis of numerical stability and convergence of nonlinear partial differential equations is done by linearization, in general [4]. Here linearization has been done on a *local* basis. This means for numerical analysis of stability properties of time-dependent NAVIER-STOKES' equation at nth time step, linearization of the nonlinear finite difference model was done by using information obtained at the $(n - 1)$th time step. This, however, should not affect the results obtained previously [1, 4], provided the assumptions in those previous works are accepted. Here, based upon computational evidence, some modifications of those assumptions were done. Errors originating from truncation of TAYLORS' series for finite difference modelling and those originating from inaccurate representations of boundary conditions should have significant effects upon the finite difference solution of NAVIER-STOKES' equations. In the first part of our study [2] an analysis of errors originating from initial-boundary conditions with regard to the explicit finite difference solution of NAVIER-STOKES' equation was done. This article consists of the second part of our study dealing with the implicit finite difference solution of the same mathematical model. Since, in any computational problem it is somewhat wise to undertake the worst possible conditions for solution, an attempt has been made in this work to seek for such conditions at the end of our analysis. However, we will see that no corrections of these conditions are possible without making significant alterations of the numerical model.

2. A Flow Model

Let us consider the following coupled partial differential equations:

$$\psi_{xx} + \psi_{yy} = -\zeta\,, \tag{2.1}$$

$$u = \psi_y\,, \qquad v = -\psi_x\,, \tag{2.2}$$

$$\zeta_t + u\zeta_x + v\zeta_y = \nu(\zeta_{xx} + \zeta_{yy}) \tag{2.3}$$

where, ψ = stream function; u, v = velocity components along the x- and y-axes respectively, ζ = vorticity, and ν = kinematic coefficient of viscosity.

These equations represent certain viscous fluid flow, the exact nature of which is determined by the initial-boundary conditions. In the absence of appropriate closed form solution, yet to be found, these equations are generally solved by numerical methods. It has been found in the past that implicit finite difference methods are quite efficient to solve such problems [1, 3, 4]. Instabilities arising from explicit finite difference methods involving marching forward scheme [2] are not in general present in the implicit numerical solution even when larger time steps are used. However, this may not hold true if errors originating from the boundary conditions are significantly large. In this article we will see that when these errors are taken into consideration, implicit finite difference solution of (2.1), (2.2), (2.3) subject to appropriate initial-boundary conditions may not be as effective as it is often expected to be.

Origin of truncation errors in finite difference analysis of differential equations in general is well known. But to analyze the net effect of errors originating from the boundary conditions one must consider a specific problem related to viscous flow. In the error equation (4) in [2], $E^n_{i,j}$, which represents the net effect of errors originating solely from the boundary conditions at a node (x_i, y_j) and at time step $t = t'_n$ is assumed to exist rather intuitively. Here, one may like to see its presence through more detailed analysis. So, we consider unsteady, laminar, viscous fluid flow past an infinite flate plate occupying the horizontal plane. The motion is two dimensional and U_∞ is the velocity of the free stream along the positive direction of the x-axis.

The outer-boundary conditions are:

$$\psi = \psi_0 + U_\infty y\,, \qquad u = U_\infty\,, \quad v = 0\,, \quad \zeta = 0 \tag{2.4}$$

and, on the body surface:

$$\psi = \psi_0 = \text{constant}\,, \qquad u = v = 0 \qquad \zeta = \zeta_0(x) = -(\psi_{xx} + \psi_{yy})\,. \tag{2.5}$$

Assuming the motion started impulsively from rest, at $t = 0$, $u = U_\infty$, $\zeta = 0$.

We note that the equation (2.3) for ζ is nonlinear and the boundary condition for ζ in (2.5) involve derivatives. These derivatives are to be approximated by finite differences and hence, just like discretization errors in the flow field, discretization errors will be introduced on the boundary.

3. Implicit Discretization

Let us assume that the model under consideration is mathematically well-posed. We also assume that in the domain of integration (i) ζ is continuously differentiable at least twice with respect to t and four times with respect to x and y, and (ii) ψ is continuously differentiable at least four times with respect to x and y. Approximating the space derivatives by central differences and time derivative by a two-point backward difference formula we get an implicit discretized model as follows:

$$\psi^n_{i,j} = \alpha(\psi^n_{i+1,j} + \psi^n_{i-1,j}) + \beta(\psi^n_{i,j+1} + \psi^n_{i,j-1}) + \gamma\zeta^n_{i,j} + \gamma T^n_{i,j}\,, \tag{3.1}$$

$$\gamma = 1/\{2(h^{-2} + l^{-2})\}\,; \qquad \alpha = \gamma h^{-2}; \qquad \beta = \gamma l^{-2}\,, \tag{3.2}$$

$$u^n_{i,j} = (\psi^n_{i,j+1} - \psi^n_{i,j-1})/2l + \delta_1\,, \tag{3.3}$$

$$v^n_{i,j} = -(\psi^n_{i+1,j} - \psi^n_{i-1,j})/2h + \delta_2\,, \tag{3.4}$$

$$(1 + 2r_1 + 2r_2)\,\zeta^n_{i,j} + (\alpha u^n_{i,j} - r_1)\,\zeta^n_{i+1,j} - (au^n_{i,j} + r_1)\,\zeta^n_{i-1,j} + + (bv^n_{i,j} - r_2)\,\zeta^n_{i,j+1} - (bv^n_{i,j} + r_2)\,\zeta^n_{i,j-1} = \zeta^{n-1}_{i,j} + \tau^n_{i,j} \tag{3.5}$$

where, $k = \Delta t$, $h = \Delta x$, $l = \Delta y$, $r_1 = \nu k/h^2$, $r_2 = \nu k/l^2$, $a = k/2h$, $b = k/2l$, $\zeta^n_{i,j} = \zeta(x_i, y_j, t_n)$; Δx, Δy being the mesh sizes and Δt being the time step.

$T^n_{i,j}$ = terms of the $O(h^2, l^2)$, $\qquad \delta_1$ = terms of the order (l^2),

$\tau^n_{i,j}$ = terms of the $O(k^2, kh^2, kl^2)$, $\qquad \delta_2$ = terms of the order (h^2).

These are known as truncation errors and due to the assumptions of smoothness of ψ and ζ, as $h \to 0$, $l \to 0$, $k \to 0$, $T_{i,j} \to 0$, δ_1, $\delta_2 \to 0$ and $\tau^n_{i,j} \to 0$. By virtue of these results, we claim that the discretized model is *consistent*.

Also, in the flow field, $i = 1, 2, \ldots, I$; $j = 1, 2, \ldots, J$; $n = 1, 2, \ldots$; and x-axis is taken along the flat plate. On the body surface, $j = 0$ and at the outer boundary $j = J + 1$. Let the flow field be given by $i = 0$, $i = I + 1$ which represent lines parallel to the y-axis.

Since x-axis is along the flat plate,

$$\zeta_{i,0} = (-\psi_{yy})_{i,0} = -h^{-2}(\psi_{i,1} - 2\psi_{i,0} + \psi_{i,-1})$$

where $\psi_{i,0} = \psi_0$, and $i = 0, 1, 2, \ldots, I + 1$.

Also, since $u = 0$ on the plate, $(\psi_{i,1} - \psi_{i,-1})/(2h) = 0$, which gives $\psi_{i,1} = \psi_{i,-1}$. Hence,

$$\zeta_{i,0} = -2h^{-2}(\psi_{i,1} - \psi_0)\,. \tag{3.6}$$

This is same as equation (3—435 A) in [4].

Obviously, then, along the body as i changes, $\zeta_{i,0}$ also changes. For this problem, at the outer boundary, $j = J + 1$,

$$\psi_{i,J+1} = U_\infty y_{J+1} + \psi_0\,, \qquad \zeta_{i,J+1} = 0\,. \tag{3.7}$$

Let $P^n_{i,j}$ and $Z^n_{i,j}$ be the net stream function and net vorticity function respectively. Let $U^n_{i,j}$, $V^n_{i,j}$ be the net velocity functions along the x and y axes respectively. Let,

$$e^n_{i,j} = \zeta^n_{i,j} - Z^n_{i,j}\,, \qquad f^n_{i,j} = \psi^n_{i,j} - P^n_{i,j} \tag{3.8), (3.9}$$

where $i = 1, 2, \ldots, I$; $j = 1, 2, \ldots, J$. Equations (3.8) and (3.9) give the discretization errors in the computational flow field.

On the body surface:

$$e^n_{i,0} = \zeta^n_{i,0} - Z^n_{i,0}\,, \qquad f^n_{i,0} = \psi^n_0 - P^n_0 = 0\ \forall\, n\,, \qquad (i = 0, 1, 2, \ldots, I + 1)\,. \tag{3.10), (3.11}$$

At the outer boundary, for all time step t_n

$$e^n_{i,J+1} = \zeta^n_{i,J+1} - Z^n_{i,J+1} = 0 \tag{3.12}$$

(since $\zeta = 0$ at the outer boundary, equation (3.7)),

$$f^n_{i,J+1} = \psi^n_{i,J+1} - P^n_{i,J+1} = 0 \tag{3.13}$$

(since both $\psi^n_{i,j+1}$ and $P^n_{i,J+1}$ will be given by the first part of equation (3.7)) $i = 0, 1, 2, \dots, I + 1$. For the *particular* model under consideration, from the initial conditions:

$$e^0_{i,j} = f^0_{i,j} = 0\,, \qquad i = 0, 1, 2, \dots, I + 1; \qquad j = 0, 1, 2, \dots, J + 1\,. \tag{3.14}$$

Thus, in this particular model discretization errors due to initial-boundary conditions are limited to the equation (3.10) only. Certainly this is not true in general. The errors originating from the outer boundary conditions (3.12) should have also a strong impact upon the numerical solution when the computational wake behind a solid body reaches the outer boundary of the flow field. This was verified in [11], where we considered a two-dimensional viscous flow past a circular cylinder. In this work we will develop a locally linearized stability analysis using matrix formulation in such a way that errors originating both from the conditions on the body surface and those from the outer boundary may be analyzed.

If we replace $\psi^n_{i,j}$ by $P^n_{i,j}$ in (3.1) and $\zeta^n_{i,j}$ by $Z^n_{i,j}$, $u^n_{i,j}$ by $U^n_{i,j}$, $v^n_{i,j}$ by $V^n_{i,j}$, in (3.5) and drop the truncation errors $T^n_{i,j}$ and $\tau^n_{i,j}$ from the equation, we get the implicit finite difference approximations of (2.1) and (2.3) as:

$$P^n_{i,j} = \alpha(P^n_{i+1,j} + P^n_{i-1,j}) + \beta(P^n_{i,j+1} + P^n_{i,j-1}) + \gamma Z^n_{i,j}\,, \tag{3.15}$$

$$(1 + 2r_1 + 2r_2)\, Z^n_{i,j} + (aU^n_{i,j} - r_1)\, Z^n_{i+1,j} - (aU^n_{i,j} + r_1)\, Z^n_{i-1,j} + (bV^n_{i,j} - r_2)\, Z^n_{i,j+1} - (bV^n_{i,j} + r_2)\, Z^n_{i,j-1} = Z^{n-1}_{i,j}\,. \tag{3.16}$$

These two equations along with the finite difference boundary conditions for $P^n_{i,j}$ and $Z^n_{i,j}$ given by (3.6) and (3.7) are to be solved numerically. This coupled system is generally solved as follows: at each t_n, we make a guess for $Z^n_{i,j}$ (usually the guess is given by the $Z^{n-1}_{i,j}$), solve (3.15) for $P^n_{i,j}$ compute $U^n_{i,j}$ $V^n_{i,j}$ (using net function equations corresponding to (3.3) and (3.4)), and then compute $Z^n_{i,1}$ from (3.16) which is then used in (3.15) for solution of $P^n_{i,j}$ at the next iteration.

Thus discretization errors originating from (3.15) will affect the numerical solution of (3.16) and vice versa. Also, one can see that in order to solve for $Z^n_{i,1}$ at any i and $n > 0$, we need to use $Z^n_{i,0}$ in the equation (3.16), which contains $e^n_{i,0}$ (3.10).

4. Matrix Equations for $P^n_{i,j}$ and $Z^n_{i,j}$

Linearized stability analysis of the coupled equations (3.15) and (3.16) require matrix formulation of these equations. In order to do this, we follow [3] and [1] and define the following I-order square matrices (at each time step t_n):

In the field:

$$\left.\begin{aligned} (P_j)_{i,s} &= \delta_{i,s} P_{i,j}\,, & (Z_j)_{i,s} &= \delta_{i,s} Z_{i,j}\,,\\ (I_I)_{i,s} &= \delta_{i,s}\,, & (L_I)_{i,s} &= \delta_{i-1,s}\,,\\ (U_j)_{i,s} &= \delta_{i,s} U_{i,j}\,, & (V_j)_{i,s} &= \delta_{i,s} V_{i,j}\,. \end{aligned}\right\} \tag{4.1}$$

On the Boundary:

$$(C_j)_{i,s} = \delta_{i,1}\delta_{1,s} Z_{0,j}; \qquad (D_j)_{i,s} = \delta_{i,I}\delta_{I,s} Z_{I+1,j}\,, \tag{4.3}$$

$$(G_j)_{i,s} = \delta_{i,1}\delta_{1,s} P_{0,j}; \qquad (H_j)_{i,s} = \delta_{i,I}\delta_{I,s} P_{I+1,j} \tag{4.2}$$

where,

$$\begin{aligned} \delta_{i,s} &= 1 \quad \text{if} \quad i = s\,, \qquad i = 1, 2, \dots, I; \qquad s = 1, 2, \dots, I\,.\\ &= 0 \quad \text{if} \quad i \neq s\,, \end{aligned} \tag{4.3}$$

Obviously,

$$(L^T_I)_{i,s} = \delta_{i+1,s}\,. \tag{4.4}$$

J-order Square Matrices with the I-order square matrices as elements:

In the Field:

$$\left.\begin{aligned} (P)_{j,s} &= \delta_{j,s} P_j; & (Z)_{j,s} &= \delta_{j,s} Z_j\,,\\ (U)_{j,s} &= \delta_{j,s} U_j; & (V)_{j,s} &= \delta_{j,s} V_j\,,\\ (I)_{j,s} &= \delta_{j,s} I_I; & (L)_{j,s} &= \delta_{j,s} L_I; \quad (B)_{j,s} = \delta_{j-1,s} I_I\,. \end{aligned}\right\} \tag{4.5}$$

On the Boundary:

$$(C)_{j,s} = \delta_{j,s} C_j; \qquad (D)_{j,s} = \delta_{j,s} D_j; \qquad (G)_{j,s} = \delta_{j,s} G_j; \qquad (H)_{j,s} = \delta_{j,s} H_j\,, \tag{4.6}$$

$$(M)_{j,s} = \delta_{j,1}\delta_{1,s} Z_0; \qquad (N)_{j,s} = \delta_{j,J}\delta_{J,s} Z_{J+1}; \qquad (R)_{j,s} = \delta_{j,1}\delta_{1,s} P_0; \qquad (S)_{j,s} = \delta_{j,J}\delta_{J,s} P_{J+1}\,. \tag{4.7}$$

Obviously,

$$(B)^T_{j,s} = \delta_{j+1,s} I_I\,, \qquad j = 1, 2, \dots, J; \qquad s = 1, 2, \dots, J\,. \tag{4.8}$$

In [3], similar matrices have been used to represent (3.15) and (3.16) in matrix forms. An explanation for this was also given in [1].
Finally, we define the following column matrices:

$$\begin{array}{ll} I\text{-order} & J\text{-order (with } I\text{-order column matrices as elements)} \\ (P_j)_i = P_{i,j} & (P)_j = P_j \\ (Z_j)_i = Z_{i,j} & (Z)_j = Z_j \\ (U_j)_i = U_{i,j} & (U)_j = U_j \\ (V_j)_i = V_{i,j} & (V)_j = V_j \\ (C_j)_i = \delta_{i,1} Z_{0,j} & (C)_j = C_j \\ (D_j)_i = \delta_{i,I+1} Z_{I+1,j} & (D)_j = D_j \\ (G_j)_i = \delta_{i,1} P_{0,j} & (G)_j = G_j \\ (H_j)_i = \delta_{i,I} P_{I+1,j} & (H)_j = H_j \\ (i = 1, 2, \dots, I) & (M)_j = \delta_{j,1} Z_0 \\ & (N)_j = \delta_{j,J} Z_{J+1} \\ & (R)_j = \delta_{j,1} P_0 \\ & (S)_j = \delta_{j,J} P_{J+1} \\ & (j = 1, 2, \dots, J) \end{array} \tag{4.9}$$

The matrix representations of (3.15) and (3.16) along with the boundary conditions at each t_n are:

$$P^n = \alpha\{(L^T P + H) + (LP + G)\} + \beta\{(B^T P + S) + (BP + R)\}^n + \gamma Z^n, \tag{4.10}$$

$$\{(1 + 2r_1 + 2r_2) Z + (aU^n - r_1 \hat{I}) (L^T Z + D) - (aU^n + r_1 \hat{I}) (LZ + C) + (bV^n - r_2 \hat{I}) (B^T Z + N) - (bV^n + r_2 \hat{I}) (BZ + M)\}^n = Z^{n-1} \tag{4.11}$$

where $\hat{I} = (1\ 1\ \dots\ 1)^T$, with IJ number of 1's. These two coupled equations show the presence of boundary conditions explicitly and hence any matrix analysis of these equations will be affected by them.

5. Coupled Error Equations

To formulate the linearized coupled matrix equations for discretization errors we replace U^n by U_{n-1} and V^n by V_{n-1} where U_{n-1}, V_{n-1} are some representative velocities at the $(n-1)$th time step. If we now consider the matrices (4.1)—(4.9) and represent $\psi_{i,j}^n$, $\zeta_{i,j}^n$, $T_{i,j}^n$, $\tau_{i,j}^n$, $e_{i,j}^n$, and $f_{i,j}^n$ accordingly, then subtracting (4.10) from the corresponding equation for ψ, and (4.11) from the corresponding equation for ζ and applying the boundary conditions (3.10)—(3.13) we get:

$$f^n = \alpha(L^T + L) f^n + \beta(B^T + B) f^n + \gamma e^n + \gamma T^n \tag{5.1}$$

and using, $W^n = \delta_{j,1}(\zeta_0^n - Z_0^n) =$ error from boundary condition,

$$(1 + 2r_1 + 2r_2) e^n + (aU_{n-1} - r_1)(L^T e^n) - (aU_{n-1} + r_1)(Le^n) - (bV_{n-1} - r_2)(B^T e^n) - (bV_{n-1} + r_2)(Be^n + W^n) = e^{n-1} + \tau^n. \tag{5.2}$$

These two equations may now be expressed as:

$$\Theta \cdot f^n = \gamma(e^n + T^n), \qquad \Phi_n e^n = e^{n-1} + E^n + \tau^n \tag{5.3), (5.4}$$

where,

$$\Theta = I - \alpha(L^T + L) - \beta(B^T + B), \tag{5.5}$$

$$\Phi_n = (1 + 2r_1 + 2r_2) I - r_1(L^T + L) + aU_{n-1}(L^T - L) - r_2(B^T + B) + bV_{n-1}(B^T - B), \tag{5.6}$$

$$E^n = (bV_{n-1} + r_2) W^n. \tag{5.7}$$

E^n is generated from the approximated boundary conditions for ζ. One may see this in the element form by subtracting (3.16) from (3.5) and using $j = 1$.

If $\xi_{p,q}^n$ and $\eta_{p,q}$ are eigenvalues of Φ_n and Θ respectively,

$$\xi_{p,q}^n = 1 + 2r_1 + 2r_2 + 2\{r_1^2 - (aU_{n-1})^2\}^{1/2} \cos\frac{\pi p}{I+1} + 2\{r_2^2 - (bV_{n-1})^2\}^{1/2} \cos\frac{\pi q}{J+1} \tag{5.8}$$

and,

$$\eta_{p,q} = 1 + 2\alpha \cos\frac{\pi p}{I+1} + 2\beta \cos\frac{\pi q}{J+1}, \qquad p = 1, 2, \dots, I; \qquad q = 1, 2, \dots, J. \quad [1, 3]. \tag{5.9}$$

From the definition of r_1, r_2, a, b, α, β, it is clear that $\xi_{p,q}^n \neq 0$, $\eta_{p,q} \neq 0$ for all p, q, n. Thus we have,

Theorem 1: *Θ and Φ_n are nonsingular for all n.*

If

$$R_x^{n-1} = \frac{hU^{n-1}}{\nu}, \qquad R_y^{n-1} = \frac{lV_{n-1}}{\nu},$$

then

$$\xi^n_{p,q} = 1 + 2r_1 + 2r_2 + r_1\{4 - (R_x^{n-1})^2\}^{1/2} \cos \frac{\pi p}{I+1} + r_2\{4 - (R_y^{n-1})^2\}^{1/2} \cos \frac{\pi q}{J+1}. \tag{5.10}$$

It can be proved that for all R_x^{n-1} and R_y^{n-1} and independent of n

$$|\xi^n_{p,q}| > \lambda > 1 \tag{5.11}$$

Hence,

$$\varrho(\Phi_n) = \text{spectral radius of } \Phi_n > \lambda > 1\,. \tag{5.12}$$

Let

$$Q_n = \Phi_n^{-1}\,. \tag{5.13}$$

Then:

Theorem 2: *Q_n is a convergent matrix.*

Proof:

$$|\lambda_{p,q}(Q_n)| = |\lambda_{p,q}(\Phi_n^{-1})| = \frac{1}{|\xi_{p,q}|} < 1 \;\forall_{p,q} \qquad (\lambda_{p,q} = \text{eigenvalues of } Q_n)\,. \tag{5.14}$$

Hence the theorem is proved.

Definition 1: *The implicit finite difference equation* (3.16) *is said to be numerically stable if Φ_n is nonsingular and Φ_n^{-1} is convergent.*

By virtue of theorems 1 and 2 we see that (3.16) is numerically stable. In the next section we will see that the above definition of numerical stability implies that if $n \to \infty$, then $e^n \to 0$ if we disregard the terms E^n and τ^n in (5.4). Therefore, it follows from (5.3), since Θ^{-1} exists and is bounded (independent of n) as $e^n \to 0$, $f^n \to 0$ if the truncation error T^n is neglected. Thus for the coupled finite difference equations (3.15) and (3.16) stability of the vorticity equation will imply the stability of the stream function equation. The converse is also true from (5.3). However, whereas, the growth of f^n can not be explicitly seen in (5.3), that of e^n can be recorded mathematically in (5.4). Hence the overall stability of the coupled system is strongly affected by the error originating from the non-linear vorticity transport equation. In our particular model, E^n contains errors originating from the body surface alone. However, the analysis presented in this article may be applied to problems where errors could also originate from the outer boundary.

6. Propagations of E^n and τ^n

From (5.4) and (5.13) we have:

$$e^n = Q_n e^{n-1} + Q_n(E^n + \tau^n)\,. \tag{6.1}$$

Applying (6.1) recurently we get:

$$e^n = Q_n Q_{n-1} \cdots Q_1 e^0 + Q_n(E^n + \tau^n) + Q_n Q_{n-1}(E^{n-1} + \tau^{n-1}) + \dots + Q_n Q_{n-1} \cdots Q_1(E^1 + \tau^1)\,. \tag{6.2}$$

Now from (5.12) and (5.14)

$$\varrho(Q_n) < \lambda^{-1} < 1\,, \quad \forall n\,. \tag{6.3}$$

Assuming U_{n-1} and V_{n-1} having same characteristic features $\forall n$ matrices Q_n are all alike. Thus (6.3) suggests that there exists a norm $\|\cdot\|_m$ such that

$$\|Q_n\|_m < \lambda^{-1} < 1 \;\forall n\,.$$

Therefore we have,

Theorem 3: $\lim_{n\to\infty} Q_n Q_{n-1} \cdots Q_1 = O\,.$ (6.4)

Proof:

$$\|Q_n Q_{n-1} \cdots Q_1\|_m \leqq \|Q_n\|_m \|Q_{n-1}\|_m \cdots \|Q_1\|_m \leqq \lambda^{-n}\,.$$

Since $\lambda^{-1} < 1$, as $n \to \infty$, $\|Q_n Q_{n-1} \cdots Q_1\| \to 0$. Now, $\|A\| = 0$ if $A = O$. Hence the theorem follows.

For the particular model under consideration e^0 = discretization error due to initial conditions = 0. But, in general, this is not the case. Still, using (6.4) in (6.2) we note that as $n \to \infty$, error due to initial conditions will be damped out. Thus, for some $n \geqq N$, where N is arbitrarily large,

$$e^N = Q_N(E^N + \tau^N) + Q_N Q_{N-1}(E^{N-1} + \tau^{N-1}) + \dots + Q_N Q_{N-1} \cdots Q_1(E^1 + \tau^1)\,. \tag{6.5}$$

This shows that by applying the so-called stability criteria (Definition 1), we have simply damped out the errors due to initial conditions but those originating from the boundary conditions added to the truncation errors (however small) are not damped out and may have a strong impact upon the discretization errors.

Considering a worst possible case, if we assume that in (6.5) terms are all positive,

$$Q_N(E^N + \tau^N) < e^N\,.$$

Thus for the m-norm,

$$\|E^N + \tau^N\|_m < \delta_1 \|e^N\|_m \tag{6.6}$$

where

$$\delta_1 = \max_N ||Q_N^{-1}||_m . \tag{6.7}$$

Also, from (6.5), for the same norm,

$$||e^N||_m \leqq \frac{\delta_2}{1-\delta_2}\,\theta \tag{6.8}$$

where,

$$\theta = \max_N ||E^N + \tau^N||_m , \qquad \delta_2 = \max_N ||Q_N||_m . \qquad (6.9), (6.10)$$

Thus, from (6.6) and (6.8)

$$\delta_1^{-1}\, ||E^N + \tau^N||_m < ||e^N||_m \leqq \delta_2\theta/(1-\delta_2) . \tag{6.11}$$

This inequality shows that in the worst possible case, discretization errors are controlled by the combined errors due to truncation and inaccurate boundary conditions.

If we further assume that: (i) Errors due to boundary conditions are much larger than those due to truncation, and, (ii) boundary conditions are independent of time, then from (6.11)

$$\delta_1^{-1}\, ||E^N||_m < ||e^N||_m \leqq \delta_2 \max_N ||E^N||/(1-\delta_2) . \tag{6.12}$$

This inequality shows that errors due to boundary conditions could totally dominate the discretization errors.

5. Conclusion

In our present analysis we have applied a local linearization by replacing $u_{i,j}^n$, $v_{i,j}^n$ by representative velocities U_{n-1} and V_{n-1} respectively. Thus equation (6.2) gives a representative growth of discretization errors which depends upon the propagation of truncation error and error due to erronous finite difference representation of boundary condition. Computationally, it is known that these two errors do have a strong impact upon numerical solution of Navier-Stokes' equations.

A question, which is quite natural, was raised [7] regarding corrections of these errors. It is obvious that if we use the vorticity finite difference model given by (3.16) these errors are possibly unavoidable. However, following "compact" difference schemes, advocated by Collatz [8], finite difference models having higher oder accuracies may be obtained. Hirsh [9] presented fourth-order accurate methods to solve Navier-Stokes equation for incompressible flows. Representations of boundary conditions for flows past curved bodies, by using rectangular coordinates introduce significant errors in the difference schemes. Such errors may be handled by using a body-fitted curvilinear co-ordinate system as discussed by Thompson [10]. However, in order to eliminate these errors originating from the boundary, a significantly more complicated mathematical model was constructed whose numerical solution seems to be rather complicated.

6. Acknowledgement

This work was supported by the Minna-James-Heinemann-Stiftung Foundation, Hannover, Germany and conducted at von Kármán Institute for Fluid Dynamics, Rhone-St-Genese, Belgium. Suggestions from the reviewer are thankfully acknowledged.

References

1 Dey, S. K., Finite Difference Analysis of Time-dependent Fluid Flow at Large Reynold's Numbers, (Accepted for publication) Bulletin of Calcutta Mathematical Society.
2 Dey, S. K., Effects of Initial Boundary Conditions on Explizit Numerical Solution of Navier-Stokes Equation, ZAMM, 222—223, Band 59, Heft 5, 1979.
3 Thompson, J. F., Optimized Acceleration of Convergence of an Implicit Numerical Solution of the Time-dependent Navier-Stokes Equations, AIAA J., Vol. 7, No. 11, Nov. 1969.
4 Roache, P. J., Computational Fluid Dynamics, Hermosa Publishers, 1976.
5 Ames, W. F., Numerical Methods for Partial Differential Equations, Barnes and Noble Inc. New York, 1969.
6 Dey, S. K., Numerical Stability of Vorticity Transport Equations in a Laminar Flow, presented at the International Congress of Mathematicians, Helsinki, Finland, 1978.
7 Hirsch, Ch., Aeronautical Eng. Dept. Vrije Universiteit Brussels (private communications).
8 Collatz, L., Numerical Treatment of Differential Equations, Springer Verlag, Berlin, 1960.
9 Hirsh, R. S., Higher Order Accurate Difference Solutions of Fluid Mechanics Problems by a Compact Differencing Technique, J. Computational Physics, Vol. 19, 1975.
10 Thompson, J. F., Numerical Solution of Flow Problems Using Body-Fitted Coordinate System, Lecture Series 1978—4, Computational Fluid Dynamics, von-Kármán-Institute for Fluid Dynamics, March, 1978.
11 Dey, S. K., Computer Programming of Fluid Flow Phenomena Past a Circular Cylinder, COED Trans. American Society of Electrical Engineers, Vol. VIII, No. 12, Dec. 1976.

Eingereicht am 8. 10. 1979, in revidierter Fassung am 2. 5. 1980

Anschrift: S. K. Dey, Professor, Department of Mathematics, Eastern Illinois University, Charleston, Illinois 61920, USA.

ZAMM 61, 105 – 114 (1981)

B. P. Sommeijer / P. J. van der Houwen

On the Economization of Stabilized Runge-Kutta Methods with Applications to Parabolic Initial Value Problems

A modification of Runge-Kutta methods is analysed which leads for important classes of parabolic differential equations to a considerable reduction of the computational effort. The main characteristic of the modified methods is the replacement of the right hand side function of the differential equation by a "cheaper" function with roughly the same Jacobian matrix. Numerical experiments are reported and the results are compared with the results obtained by the unmodified Runge-Kutta method and by an ADI method.

Eine Modifikation von Runge-Kutta-Methoden wird analysiert, die für wichtige Klassen parabolischer Differentialgleichungen zu einer beträchtlichen Reduktion des Rechenaufwandes führt. Hauptcharakteristik der modifizierten Methoden ist die Ersetzung der Funktion auf der rechten Seite der Differentialgleichung durch eine „billigere" Funktion mit grob gesprochen derselben Jacobi-Matrix. Über numerische Experimente wird berichtet, und die Ergebnisse werden mit denen verglichen, die man durch die nichtmodifizierte Runge-Kutta-Methode und durch eine ADI-Methode erhält.

Анализируется модификация методов типа Рунге-Кутта ведущая для важных классов параболических дифференциальных уравнении на значительное упрощение вычислительных затрат. Главная характеристика модифицированных методов состоит в том, что замещается функцию на правом стороне дифференциального уравнения «Более дешевую» функцию имеющую приблизительно та самая матрица типа Якоби. Доносятся о вылислительные эксперименты и сравняются результаты с результатами полученными немодифицированным методам типа Рунге-Кутта и некоторым методам типа АДИ.

1. Introduction

Suppose we are given a time-dependent partial differential equation defined on a domain Ω with boundary $\partial\Omega$ in the space of the space variable $\boldsymbol{x}$. By applying the method of lines, that is by replacing $\Omega \cup \partial\Omega$ by a set of grid points $\Gamma_h \cup \partial\Gamma_h$, h referring to the coarseness of the grid, such a partial differential equation (p.d.e.) is converted into a system of ordinary differential equations which frequently is of the form

$$\frac{d\boldsymbol{y}}{dt} = \boldsymbol{f}(t, \boldsymbol{y} + \boldsymbol{b}), \qquad \boldsymbol{b} = \boldsymbol{g}(t, \boldsymbol{y}). \tag{1.1}$$

Here, $\boldsymbol{y}$ denotes an approximation to the solution of the p.d.e. at the internal grid points Γ_h, being zero at the boundary points $\partial\Gamma_h$, and $\boldsymbol{b}$ denotes an approximation to the solution at the boundary points $\partial\Gamma_h$ being zero at Γ_h. The functions $\boldsymbol{f}$ and $\boldsymbol{g}$ are given, having zero-components at $\partial\Gamma_h$ and Γ_h, respectively. The function $\boldsymbol{g}$ originates from the boundary conditions and expresses the boundary values in terms of the solution at the internal grid points Γ_h.

In this paper we study a modification of *stabilized* Runge-Kutta *methods* (RK methods) for the solution of the initial value problem for equation (1.1). We shall call an RK method *stabilized* when extra function evaluations are added in order to increase the stability boundary. Large stability boundaries are desirable when the spectral radius of the Jacobian of $\boldsymbol{f}$ with respect to $\boldsymbol{y}$ is large which is just the situation when (1.1) originates from a p.d.e.. In the case of parabolic p.d.e.'s, stabilized RK methods have been constructed with a (real) stability boundary $\beta \cong c_p m^2$ where m is the number of function evaluations per step and c_p is a constant depending on the order p of the RK method ($c_1 \cong 1.93$, $c_2 \cong .80$, $c_3 \cong .48$ [3]). The *maximal, stable integration step* of these methods is given by

$$\tau_{\max} = \frac{\beta}{\sigma(J_n)} \cong \frac{c_p m^2}{\sigma(J_n)}, \tag{1.2}$$

where $\sigma(J_n)$ denotes the spectral radius of the Jacobian matrix

$$J_n = \left.\frac{\partial \boldsymbol{f}(t_n, I\boldsymbol{v} + \boldsymbol{g}(t_n, \boldsymbol{v}))}{\partial \boldsymbol{v}}\right|_{\boldsymbol{v}=\boldsymbol{y}_n}, \tag{1.3}$$

$\boldsymbol{y}_n$ denoting the numerical approximation at $t = t_n$ and I being the operator which replaces the boundary components of a vector $\boldsymbol{v}$ by zeros; thus J_n only contains derivatives corresponding to the gridpoints of Γ_h. Although stabilized RK methods are available (for semi-discrete parabolic equations) in which the number of stages can be chosen arbitrarily large without danger of internal instabilities [4], these methods generally are more expensive in terms of right hand side evaluations than implicit or partially implicit integration methods. It is often possible, however, to reduce the work per integration step by exploiting the fact that a stabilized RK method usually contains a small number of $\boldsymbol{f}$-evaluations which ensure the order of accuracy and a large number of $\boldsymbol{f}$-evaluations which take care of the stability of the scheme. This means that these last $\boldsymbol{f}$-evaluations may be replaced by a function $\boldsymbol{f}^*(t, \boldsymbol{y})$ without affecting the *order of accuracy*. In order to obtain the same stability regions one should choose $\boldsymbol{f}^*$ such that its Jacobian matrix is a first order approximation in τ to the original Jacobian matrix. When the effort involved to compute the set of $\boldsymbol{f}^*$-vectors within an integration step is much less than the computational effort involved to evaluate the replaced $\boldsymbol{f}$-vectors, then computing time can be saved by this simple modification of the RK method. We call such methods *modified* Runge-Kutta *methods*.

As an example, consider the function

$$f^*(t, y) = J_n y\,, \qquad t_n \leqq t \leqq t_{n+1}\,, \tag{1.4}$$

which is a plausible choice in cases of linear problems where J_n is directly available. In [2] a few experiments were performed with this modification. It turned out that stability is preserved but the accuracy is considerably reduced. Here we consider more general and at the same time more accurate f^*-functions and investigate the effect on the accuracy of the numerical solution. Numerical experiments are reported and the results are compared with those obtained by the ADI method of PEACEMAN-RACHFORD. The modified RUNGE-KUTTA method seems to be particularly advantageous for strongly non-linear problems and if the costs per integration step are about 25% (according to our experiments) of the costs of the classical RUNGE-KUTTA step.

2. Modified Runge-Kutta methods

Consider the *m-stage* RUNGE-KUTTA *formula*

$$\begin{aligned} y_{n+1}^{(0)} &= y_n\,,\\ y_{n+1}^{(j)} &= y_n + \tau \sum_{l=0}^{j-1} \lambda_{jl} f^*(t_n + \theta_l \tau, y_{n+1}^{(l)})\,, \qquad j = 1, 2, \dots, m\,, \\ y_{n+1} &= y_{n+1}^{(m)}; \qquad \theta_0 = 0\,, \quad \theta_j = \sum_{l=0}^{j-1} \lambda_{jl}\,, \qquad j = 1, 2, \dots, m-1\,. \end{aligned} \tag{2.1}$$

If the function $f^*(t, y)$ is identical to the right hand side function $f(t, y + g(t, y))$ this scheme represents a classical RUNGE-KUTTA method for equation (1.1). For the parameters λ_{jl} we choose the values which define an appropriate classical stabilized RK method (cf. [4]). A few important schemes are specified in section 3.2. These are characterized by their limited storage requirements, an important feature in view of our purpose to apply them to the large systems arising from the space-discretization of partial differential equations (see also [5]). The formula (2.1) together with a prescribed function f^* define a *modified* RUNGE-KUTTA *method* (MRK *method*).

In this section the conditions are derived under which the modified formula has the same stability region and order of accuracy as the generating classical RK formula.

2.1. Stability

The stability region of the MRK method (2.1) is identical to that of the generating RK method because the stability polynomials are identical (the stability polynomial $R_m(z)$ is defined by the relation $y_{n+1} = R_m(\tau\lambda)\, y_n$ obtained by applying the method to the equation $y' = \lambda y$). In our case this polynomial is generated by the recurrence relation

$$R_0(z) = 1\,, \qquad R_j(z) = 1 + \sum_{l=0}^{j-1} \lambda_{jl} z R_l(z)\,, \qquad j = 1, 2, \dots, m\,. \tag{2.2}$$

When $J_n^* = \partial f^*/\partial y$ has negative eigenvalues λ then the modified scheme is called *stable* if (cf. (1.2))

$$\tau \leqq \frac{\beta}{\sigma(J_n^*)}\,,$$

where β is the real stability boundary of $R_m(z)$. A p-th order RK method has a stability polynomial of the form

$$R_m(z) = 1 + z + \frac{1}{2} z^2 + \dots + \frac{1}{p!} z^p + \beta_{p+1} z^{p+1} + \dots + \beta_m z^m\,, \tag{2.3}$$

where the coefficients $\beta_{p+1}, \dots, \beta_m$ are expressions in the RK parameters. For instance,

$$\beta_1 = \sum_{l=0}^{m-1} \lambda_{ml}\,, \qquad \beta_2 = \sum_{j=1}^{m-1} \lambda_{mj} \sum_{l=0}^{j-1} \lambda_{jl} = \sum_{j=1}^{m-1} \lambda_{mj}\theta_j\,. \tag{2.4}$$

Polynomials of the form (2.3) will be called *p-th order consistent.*

In this paper we consider first, second and third order consistent polynomials. In the first order case we use the polynomial [3]

$$\tilde{R}_m^{(1)}(z) = \frac{T_m\left(w_0 + \dfrac{w_0+1}{\beta} z\right)}{T_m(w_0)}\,, \qquad \beta = \frac{(w_0+1)\, T_m'(w_0)}{T_m(w_0)}\,, \qquad 0 < w_0 - 1 \ll 1\,. \tag{2.5}$$

To obtain second order we use the polynomial [4]

$$\tilde{A}_m^{(2)}(z) = a_0 + a_m T_m\left(w_0 + \frac{w_0+1}{\beta} z\right), \qquad \beta = \frac{(w_0+1)\, T_m''(w_0)}{T_m'(w_0)}\,, \qquad 0 < w_0 - 1 \ll 1\,, \tag{2.6}$$

where

$$a_m = \frac{T_m''(w_0)}{[T_m'(w_0)]^2}\,, \qquad a_0 = 1 - a_m T_m(w_0)\,.$$

Finally we use the third order consistent polynomial

$$\tilde{B}_m^{(3)}(z) = a_0 + a_l T_l(w_0 + w_1 z) + a_m T_m(w_0 + w_1 z)\,, \qquad m = 3l\,, \tag{2.7}$$

$$\beta = \frac{w_0 + 1}{w_1}, \qquad 0 < w_0 - 1 \ll 1\,,$$

where a_0, a_l, a_m and w_1 are defined by the consistency relations

$$\left[\frac{d^i}{dz^i} \tilde{B}_m^{(3)}(z)\right]_{z=0} = 1\,, \qquad i = 0, \dots, 3\,.$$

Solving these relations leads to

$$a_m = \left(\frac{T_l''(w_0)}{w_1} - \frac{T_l'(w_0)}{w_1^2}\right)\Big/\left(T_m'(w_0)\, T_l''(w_0) - T_l'(w_0)\, T_m''(w_0)\right),$$

$$a_l = \left(\frac{T_m'(w_0)}{w_1^2} - \frac{T_m''(w_0)}{w_1}\right)\Big/\left(T_m'(w_0)\, T_l''(w_0) - T_l'(w_0)\, T_m''(w_0)\right),$$

$$a_0 = 1 - a_l T_l(w_0) - a_m T_m(w_0)$$

and w_1 is the smallest root of

$$w_1^2[T_l''(w_0)\, T_m'''(w_0) - T_l'''(w_0)\, T_m''(w_0)] + w_1[T_l'''(w_0)\, T_m'(w_0) - T_m'''(w_0)\, T_l'(w_0)] + $$
$$+ [T_l'(w_0)\, T_m''(w_0) - T_m'(w_0)\, T_l''(w_0)] = 0\,.$$

In (2.5)—(2.7) T_k denotes the CHEBYSHEV polynomial of degree k.

2.2. The order of accuracy of modified RUNGE-KUTTA methods

Obviously, the function $\boldsymbol{f}^*$ in (2.1) should have some relation to the right hand side function $\boldsymbol{f}$ in order to present a consistent approximation to the equation (1.1). It will be assumed that in the interval $t_n \leqq t \leqq t_n + \tau$ this relation is of the form

$$\boldsymbol{f}^*(t_n + \theta_l\tau, \boldsymbol{y}_{n+1}^{(l)}) - \boldsymbol{f}\big(t_n + \theta_l\tau, \boldsymbol{y}_{n+1}^{(l)} + \boldsymbol{g}(t_n + \theta_l\tau, \boldsymbol{y}_{n+1}^{(l)})\big) = \delta_l\tau^q\boldsymbol{\varphi}_n + \eta_l\tau^s\boldsymbol{\psi}_n + \boldsymbol{O}(\tau^{q+1} + \tau^{s+1})\,, \tag{2.8}$$

where δ_l, η_l are scalars, q, s integers $\geqq 1$ and $\boldsymbol{\varphi}_n$, $\boldsymbol{\psi}_n$ vectors only depending on $(t_n, \boldsymbol{y}_n)$. Evidently, $\boldsymbol{\varphi}_n = \boldsymbol{\psi}_n = \boldsymbol{o}$ is $\boldsymbol{f}^* = \boldsymbol{f}$. In Section 2.3 examples are given of functions $\boldsymbol{f}^*$ satisfying (2.8). Here, we first derive the order equations.

Starting with a p-th order classical RK formula our analysis can be confined to the derivation of an expression for the difference $\boldsymbol{y}_{n+1} - \tilde{\boldsymbol{y}}_{n+1}$, where $\tilde{\boldsymbol{y}}_{n+1}$ denotes the result which would be obtained if the classical RK method is applied at $t = t_n$. Let $\tilde{\boldsymbol{y}}_{n+1}^{(j)}$ denote the intermediate vectors obtained with the classical RK method. From (2.1) and (2.8) it follows that the intermediate deviations

$$\Delta\boldsymbol{y}_{n+1}^{(j)} = \boldsymbol{y}_{n+1}^{(j)} - \tilde{\boldsymbol{y}}_{n+1}^{(j)}$$

are approximately determined by the scheme

$$\Delta\boldsymbol{y}_{n+1}^{(0)} = \boldsymbol{o}\,,$$

$$\Delta\boldsymbol{y}_{n+1}^{(j)} = \tau\sum_{l=0}^{j-1}\lambda_{jl}[J_n\Delta\boldsymbol{y}_{n+1}^{(l)} + \tau^q\delta_l\boldsymbol{\varphi}_n + \tau^s\eta_l\boldsymbol{\psi}_n + \boldsymbol{O}(\tau^{q+1} + \tau^{s+1})]\,, \qquad j = 1, 2, \dots, m\,,$$

$$\boldsymbol{y}_{n+1} - \tilde{\boldsymbol{y}}_{n+1} = \Delta\boldsymbol{y}_{n+1}^{(m)}\,.$$

From these relations it follows that the *local deviation error* is

$$\boldsymbol{y}_{n+1} - \tilde{\boldsymbol{y}}_{n+1} = \tau^{q+1}Q_m(\tau J_n)\,[\boldsymbol{\varphi}_n + \boldsymbol{O}(\tau)] + \tau^{s+1}S_m(\tau J_n)\,[\boldsymbol{\psi}_n + \boldsymbol{O}(\tau)]\,, \tag{2.9}$$

where Q_m, S_m are polynomials in τJ_n of degree $m - 1$ defined by the recurrence relations

$$\left.\begin{aligned} Q_0(z) &= 0\,, \quad Q_j(z) = \sum_{l=0}^{j-1}\lambda_{jl}[zQ_l(z) + \delta_l] \\ S_0(z) &= 0\,, \quad S_j(z) = \sum_{l=0}^{j-1}\lambda_{jl}[zS_l(z) + \eta_l] \end{aligned}\right\} \quad j = 1, 2, \dots, m\,.$$

Hence,

$$Q_m(z) = \sum_{l=0}^{m-1}\lambda_{ml}\delta_l + \sum_{j=1}^{m-1}\sum_{l=0}^{j-1}\lambda_{mj}\lambda_{jl}\delta_l z + \dots \tag{2.10}$$

and a similar expression for $S_m(z)$.

From (2.9) it is immediate that a p-th order RK method generates an MRK method of order p if either

$$\min\,(q, s) = p \tag{2.11}$$

or

$$\begin{cases} \min\,(q, s) < p\,, \\ \left.\dfrac{d^k}{dz^k}\,Q_m(z)\right|_{z=0} \cdot \boldsymbol{\varphi}_n = \boldsymbol{o}\,, \quad k = 0, \dots, p - 1 - q\,, \quad \left.\dfrac{d^l}{dz^l}\,S_m(z)\right|_{z=0} \cdot \boldsymbol{\psi}_n = \boldsymbol{O}\,, \quad l = 0, \dots, p - 1 - s\,. \end{cases} \tag{2.11'}$$

2.3. Modified right hand side functions

In this section we consider a few possibilities for choosing the modified right hand side function $f^*(t, y)$. The *order* of the modification will be defined by min (q, s) (cf. (2.9)).

2.3.1. First order modifications

We start with discussing a few examples of modified right hand side functions:

(i) Simple examples of modifications only involving the t-argument are defined by

$$f^*(t, y) = f(t_n + \theta\tau, y + g(t_n + \theta\tau, y)), \quad t_n \leqq t \leqq t_{n+1}, \quad 0 \leqq \theta \leqq 1. \tag{2.12a}$$

$$f^*(t, y) = f(t_n + \theta\tau, y + g(t, y)) \tag{2.12b}$$

These functions are of use if the time-dependency in f forms the main part of the computational effort in the evaluation of f. For instance, in the case of a separable right hand side function f, i.e.

$$f(t, y + g(t, y)) = \sum_{i=1}^{r} T_i(t)\, Y_i(y + g(t, y)), \tag{2.13}$$

T_i being matrices only depending on t and Y_i being vector functions depending on y and t, the functions (2.12a) and (2.12b) are suitable economizations if the matrices T_i require the greater part of the computational effort to evaluate f. It should be remarked however, that generally (2.12b) yields considerably more accurate results than (2.12a). This is due to the equal levels of consistency of the g and y fields at all places where they simultaneously appear ([1], see also the experiments in Section 3.3).

(ii) When in (2.13) the matrices T_i also depend on y, i.e.

$$f(t, y + g(t, y)) = \sum_{i=1}^{r} T_i(t, y)\, Y_i(y + g(t, y)), \tag{2.13'}$$

we may define

$$f^*(t, y) = \sum_{i=1}^{r} T_i(t_n + \theta\tau, y_n)\, Y_i(y + g(t, y)). \tag{2.12c}$$

(iii) As a last example we consider a modification based on the JACOBIan matrix of f:

$$f^*(t, y) = f(t_n + \theta\tau, y_n + g(t_n, y_n)) + K_\theta[y + g(t, y) - y_n - g(t_n, y_n)], \quad 0 \leqq \theta \leqq 1, \quad t_n \leqq t \leqq t_{n+1},$$

$$K_\theta = \left.\frac{\partial f(t_n + \theta\tau, v)}{\partial v}\right|_{v = y_n + g(t_n, y_n)}. \tag{2.12d}$$

This function is sort of linearization of f in the interval $[t_n, t_{n+1}]$ and is of practical value if the JACOBIan K_θ is easily obtained. Several other linearizations are possible but (2.12d) was chosen because of its property to reduce to (2.12b) for right hand side functions of the form $A(t)\,(y + g(t, y))$, $A(t)$ being a matrix operator. It should be remarked that similarly to (2.12b) the t-argument of the boundary function g occurring in (2.12d) is chosen according to the time-level of the corresponding y-field. As already observed above the accuracy is improved by this choice.

It is easily verified that the preceding examples are of the more general form

$$f^*(t, y) = F(t_n + \theta\tau, t, y_n, y), \quad t_n \leqq t \leqq t_{n+1}, \tag{2.12}$$

where θ is a parameter in the interval [0, 1] and F a function satisfying the condition

$$F(t, t, y, y) = f(t, y + g(t, y)). \tag{2.14}$$

This modification satisfies condition (2.7) with

$$\begin{aligned} q &= 1, \quad \delta_l = \theta - \theta_l, \quad \varphi_n = \left.\frac{\partial F(t, t_n, y_n, y_n)}{\partial t}\right|_{t=t_n} \\ s &= 1, \quad \eta_l = -\theta_l, \quad \psi_n = \left.\frac{\partial F(t_n, t_n, y, y_n)}{\partial y}\right|_{y=y_n} \cdot f^*(t_n, y_n). \end{aligned} \tag{2.15}$$

Thus, (2.12) and in particular (2.12a, ..., d), represents a class of first order modifications. Let p be the order of the generating RK formula, then the following theorem is immediate from (2.11), (2.15) and (2.4):

Theorem 2.1:

(a) *The method* {(2.1); (2.12)} *is of first order for all* θ *if* $p = 1$.

(b) *The methods* {(2.1); (2.12a), (2.12b), (2.12d)} *are of second order if* $p = 2$ *and*

$$\theta = \frac{\sum_{l=1}^{m-1} \lambda_{ml}\theta_l}{\sum_{l=0}^{m-1} \lambda_{ml}} = \frac{\beta_2}{\beta_1} = \frac{1}{2}. \tag{2.16}$$

2.3.2. Second order modifications

Consider the modification

$$f^*(t, y) = \alpha(t)\, F(t_n + \bar{\theta}\tau, t, y_n, y) + (1 - \alpha(t))\, F(t_n + \bar{\bar{\theta}}\tau, t, y_n, y)\,,$$

$$\alpha(t) = \frac{t - t_n - \bar{\bar{\theta}}\tau}{(\bar{\theta} - \bar{\bar{\theta}})\,\tau}\,, \qquad 0 \leq \bar{\theta} < \bar{\bar{\theta}} < 1\,, \tag{2.17}$$

where F satisfies (2.13). For F one may choose e.g. the function defined by (2.12a), ..., (2.12d). These modifications will be indicated by (2.17a), ..., (2.17d), respectively. A straightforward calculation yields that (2.7) is satisfied with

$$q = 2\,, \qquad \delta_l = \frac{1}{2}(\bar{\theta} - \theta_l)(\theta_l - \bar{\bar{\theta}})\,, \qquad \varphi_n = \left.\frac{\partial^2 F(t, t_n, y_n, y_n)}{\partial t^2}\right|_{t=t_n},$$

$$s = 1\,, \qquad \eta_l = -\theta_l\,, \qquad \psi_n = \left.\frac{\partial F(t_n, t_n, y, y_n)}{\partial y}\right|_{y=y_n} \cdot f^*(t_n, y_n)\,. \tag{2.18}$$

A second order modification is obtained if the JACOBIan matrix $L_n = \partial F(t_n, t_n, y, y_n)/\partial y$ vanishes. We then have the following theorem:

Theorem 2.2:

(a) *The method* {(2.1); (2.17)} *is of second order for all* $\bar{\theta}$ *and* $\bar{\bar{\theta}}$ *if* $p = 2$ *and if the* JACOBIan *matrix* L_n *vanishes.*

(b) *The method* {(2.1); (2.17)} *is of third order if* $p = 3$, *the matrix* L_n *vanishes and if* $\bar{\theta}$ *and* $\bar{\bar{\theta}}$ *satisfy the equation*

$$\bar{\theta}\bar{\bar{\theta}} - \tfrac{1}{2}(\bar{\theta} + \bar{\bar{\theta}}) + \tfrac{1}{3} = 0\,. \tag{2.19}$$

Proof: Part (a) is immediate from (2.11) and (2.18). Part (b) is proved by applying (2.11'), i.e. $\bar{\theta}$ and $\bar{\bar{\theta}}$ have to satisfy the equation

$$\sum_{l=0}^{m-1} \lambda_{ml}(\bar{\theta} - \theta_l)(\bar{\bar{\theta}} - \theta_l) = 0\,.$$

Since every third order RK method satisfies the conditions

$$\sum_{l=0}^{m-1} \lambda_{ml} = 1\,, \quad \sum_{l=1}^{m-1} \lambda_{ml}\theta_l = \tfrac{1}{2}\,, \quad \sum_{l=1}^{m-1} \lambda_{ml}\theta_l^2 = \tfrac{1}{3}$$

we arrive at equation (2.19). Q.e.d.

Remarks:

(i) The application of the method {(2.1); (2.17)} with non-vanishing JACOBIan matrix L_n is certainly possible and will be more accurate as $F(t_n, t_n, y, y_n)$ is smoother with respect to y.

(ii) Suitable choices of the parameters $\bar{\theta}$ and $\bar{\bar{\theta}}$ are

Case (a): $\bar{\theta} = 0$; $\quad \bar{\bar{\theta}} = 1$

$$\text{Case (b)}: \begin{cases} \bar{\theta} = 0; & \bar{\bar{\theta}} = \tfrac{2}{3} \quad \text{for } n \text{ even}\,, \\ \bar{\theta} = \tfrac{1}{3}; & \bar{\bar{\theta}} = 1 \quad \text{for } n \text{ odd}\,. \end{cases} \tag{2.19'}$$

It is easily seen that these parameter values save computational effort when we are dealing with separable right hand side functions.

2.4. The difference between the classical and modified RUNGE-KUTTA solution

In the preceding section we derived the conditions for p-th order accuracy of the modified RK formula. In actual computation, however, the MRK methods usually are less accurate than the generating RK methods (of the same order). The reason is that the error (2.8) is generally larger than the local truncation error of the generating formula although the orders in τ are equal. Therefore, it is of interest to see how the polynomials $Q_m(z)$ and $S_m(z)$ behave because these polynomials determine largely the magnitude of the deviation (2.8). Considering modifications which satisfy (2.15), we can express the polynomials $Q_m(z)$ and $S_m(z)$ in terms of the stability polynomial $R_m(z)$.

Theorem 2.3.: *If* (2.15) *is satisfied then* $Q_m(z)$ and $S_m(z)$ in (2.9) are given by

$$Q_m(z) = \theta\,\frac{R_m(z) - 1}{z} - \frac{R_m(z) - 1 - \theta_m z}{z^2}\,, \qquad S_m(z) = -\,\frac{R_m(z) - 1 - \theta_m z}{z^2}\,. \tag{2.20}$$

Proof: Substitution of $\delta_l = \theta - \theta_l$ into (2.9) yields

$$Q_0(z) = 0\,, \qquad Q_j(z) = \theta \sum_{l=0}^{j-1} \lambda_{jl} + \sum_{l=0}^{j-1} \lambda_{jl}[zQ_l(z) - \theta_l]\,.$$

Let us define the polynomials $A_j(z)$ and $B_j(z)$ by writing

$$Q_j(z) = \theta A_j(z) - B_j(z) .$$

These polynomials satisfy the recurrence relations

$$A_0(z) = 0 , \quad A_1(z) = \lambda_{10} , \quad A_j(z) = \sum_{l=0}^{j-1} \lambda_{jl}(1 + zA_l(z)) , \quad j = 2, 3, \dots , m , \tag{2.21}$$

$$B_0(z) = B_1(z) = 0 , \quad B_j(z) = \sum_{l=0}^{j-1} \lambda_{jl}(\theta_l + zB_l(z)) , \quad j = 2, 3, \dots , m . \tag{2.22}$$

It is easily verified that

$$A_j(z) = \frac{R_j(z) - 1}{z} \quad \text{and} \quad B_j(z) = \frac{R_j(z) - 1 - \theta_j z}{z^2} \tag{2.23}$$

satisfy (2.21) and (2.22), respectively, provided that $R_j(z)$ is defined by (2.2). From (2.23) the expression (2.20) for $Q_m(z)$ follows. In a similar way we find the expression for $S_m(z)$ as given by (2.20). Q.e.d.

From this theorem it follows that

$$\frac{1}{z} \leqq Q_m(z) \leqq \frac{2 + (1 - 2\theta) z}{z^2}, \quad \frac{1}{z} \leqq S_m(z) \leqq \frac{2 + z}{z^2}, \quad -\beta \leqq z \leqq 0 , \tag{2.24}$$

where we have put $\theta_m = 1$, that is we assume the formula at least first order exact. These inequalities show that $|Q_m(z)|$ and $|S_m(z)|$ are small for large $|z|$-values so that we concentrate on their behaviour (relatively) close to the origin. Again by theorem 2.3 we have

$$Q_m(z) \cong (\theta - \beta_2) + (\theta\beta_2 - \beta_3) z + (\theta\beta_3 - \beta_4) z^2 + (\theta\beta_4 - \beta_5) z^3 \tag{2.25}$$

for sufficiently small $|z|$-values. The behaviour of $S_m(z)$ follows from this approximation by putting $\theta = 0$. In our case where we choose for $R_m(z)$ either $\tilde{R}_m^{(1)}(z)$ or $\tilde{A}_m^{(2)}(z)$ as defined by (2.5) and (2.6), respectively, the coefficients $\beta_2, \dots , \beta_5$ can be shown to be almost independent of m and hence the polynomials $Q_m(z)$ and $S_m(z)$ have for all m a more or less identical behaviour near the origin. For $\theta = \beta_2$ the polynomial $|Q_m(z)|$ assumes its maximum value .07 at $z = -10$ and .20 at $z = -3.5$ in the respective cases $\tilde{R}_m^{(1)}(z)$ and $\tilde{A}_m^{(2)}(z)$. The polynomial $|S_m(z)|$ assumes its maximum value β_2 at the origin. Thus, it is expected that *first order methods will loose less accuracy when modified than second order methods.*

2.5. Boundary conditions

The last aspect of scheme (2.1) to be discussed in this paper concerns the boundary conditions $\boldsymbol{b} = \boldsymbol{g}(t, \boldsymbol{y})$. It is well-known that in algorithms for time-dependent p.d.e.'s with intermediate stages such as RK methods or splitting methods [1], the boundary values and the internal solution values should form a sufficiently smooth grid function on $\Gamma_h \cup \partial\Gamma_h$ as soon as they simultaneously appear as argument of the function $\boldsymbol{f}$. This requirement restricts the choice of the functions $\boldsymbol{f}$ and is the reason for choosing the t-arguments of $\boldsymbol{g}(t, \boldsymbol{y})$ as done in (2.12 b, c, d). We will illustrate this by the following heuristic analysis of what happens when we choose (2.12a) and $\theta = 0$.

For the sake of simplicity $\boldsymbol{g}$ will be assumed only to depend on t (i.e. DIRICHLET boundary conditions). The formula (2.1) will contain terms

$$\boldsymbol{f}(t_n, \boldsymbol{y}_{n+1}^{(l)} + \boldsymbol{g}(t_n)) = \boldsymbol{f}\Big(t_n, \boldsymbol{y}_n + \tau \sum_{i=0}^{l-1} \lambda_{li} \boldsymbol{f}(t_n, \boldsymbol{y}_{n+1}^{(i)} + \boldsymbol{g}(t_n)) + \boldsymbol{g}(t_n)\Big). \tag{2.26}$$

Let us start the integration step at $t = t_n$ with a grid function $\boldsymbol{y}_n + \boldsymbol{g}(t_n)$ which convergees to a smooth function as the grid is refined, i.e. as $h \to 0$, and τ is kept fixed. We expect that the expression (2.26) remains only bounded as $h \to 0$ if its argument $\boldsymbol{y}_{n+1}^{(l)} + \boldsymbol{g}(t_n)$ is a sufficiently smooth function on the grid $\Gamma_h \cup \partial\Gamma_h$, that is if the grid functions

$$\boldsymbol{f}(t_n, \boldsymbol{y}_{n+1}^{(i)} + \boldsymbol{g}(t_n))$$

converge to a smooth function on $\Omega \cup \partial\Omega$. However, since $\boldsymbol{f}$ has zero-components in all boundary points $\partial\Gamma_h$, this grid function converges to a discontinuous function, and therefore we cannot expect that (2.26) is bounded as $h \to 0$. The modification (2.12b) does not give this singular behaviour, because (2.26) now reads (for general boundary functions $\boldsymbol{g}(t, \boldsymbol{y})$)

$$\begin{aligned} &\boldsymbol{f}(t_n + \theta\tau, \boldsymbol{y}_{n+1}^{(l)} + \boldsymbol{g}(t_n + \theta_l\tau, \boldsymbol{y}_{n+1}^{(l)})) \\ &= \boldsymbol{f}\Big(t_n + \theta\tau, \boldsymbol{y}_n + \boldsymbol{g}(t_n, \boldsymbol{y}_n) + \tau \sum_{i=0}^{l-1} \lambda_{li} \boldsymbol{f}(t_n + \theta\tau, \boldsymbol{v}^{(i)}) + \boldsymbol{g}\Big(t_n + \theta_l\tau, \boldsymbol{y}_n + \tau \sum_{i=0}^{l-1} \lambda_{li} \boldsymbol{f}(t_n + \theta\tau, \boldsymbol{v}^{(i)})\Big) - \boldsymbol{g}(t_n, \boldsymbol{y}_n)\Big) \end{aligned} \tag{2.26'}$$

where

$$\boldsymbol{v}^{(i)} = \boldsymbol{y}_{n+1}^{(i)} + \boldsymbol{g}(t_n + \theta_i\tau, \boldsymbol{y}_{n+1}^{(i)}) .$$

By the same reasoning as above we should now require that the grid functions

$$\boldsymbol{f}(t_n + \theta\tau, \boldsymbol{v}^{(i)}) + \frac{\boldsymbol{g}\Big(t_n + \theta_l\tau, \boldsymbol{y}_n + \tau \sum_{j=0}^{l-1} \lambda_{lj} \boldsymbol{f}(t_n + \theta\tau, \boldsymbol{v}^{(j)})\Big) - \boldsymbol{g}(t_n, \boldsymbol{y}_n)}{\tau \sum_{i=0}^{l-1} \lambda_{li}}$$

are sufficiently smooth for small h-values. If the functions $\boldsymbol{f}$ and $\boldsymbol{g}$ slowly vary with their arguments, this grid function approximates the grid function $d\boldsymbol{y}/dt + d\boldsymbol{b}/dt$ at $t = t_n$ and therefore may be assumed to be sufficiently smooth.

By a similar argument we are led to the function (2.12d) instead of the usual linearization of $\boldsymbol{f}(t, \boldsymbol{y})$.

3. Numerical Experiments

The aim of our numerical experiments was to get information to what extent the results of the MRK methods differ from the results of the generating classical methods. In order to demonstrate the relevance of the modified schemes in practical problems we also present results obtained by the ADI method which is accepted in the literature as one of the more efficient integration techniques. To compare the efficiency of the MRK and ADI methods we define the *efficiency rate* of the MRK methods with respect to the ADI method: Let τ be the stepsize for which the m-stage MRK method produces an accuracy A and let $N(A)$ be the total number of $\boldsymbol{f}$-evaluations which is needed by the ADI method to produce the same accuracy A. Then $\tau N(A)/m$ will be called the efficiency rate of the MRK method corresponding to the integration step τ. This value indicates the fraction to which the computational effort of the generating RK method should be reduced in order to make its modification as efficient as the ADI method. The rate of efficiency of an MRK method provides insight in its practical relevance. Of course, one should only apply the modification when a substantial reduction of computing time can be achieved. For the sake of illustration however this is not important and therefore we have chosen relatively simple test problems.

3.1. The initial-boundary value problems

The following two equations were chosen:

$$u_t = u_{x_1x_1} + u_{x_2x_2} - e^{-t}(x_1^2 + x_2^2 + 4), \qquad 0 \leqq t \leqq 1, \tag{3.1}$$

and

$$u_t = \frac{x_1 + x_2}{2(1+t)}\,[(u^3)_{x_1x_1} + (u^3)_{x_2x_2}] + \pi(x_1 + x_2)\cos(2\pi t) - \frac{3(x_1 + x_2)^2}{4(1+t)}\sin^3(2\pi t), \qquad 0 \leqq t \leqq 1. \tag{3.2}$$

The functions

$$u(t, x_1, x_2) = 1 + e^{-t}(x_1^2 + x_2^2) \tag{3.3}$$

and

$$u(t, x_1, x_2) = \sin(2\pi t)\,(x_1 + x_2)/2 \tag{3.4}$$

satisfy (3.1) and (3.2), respectively. Both equations were considered on the unit square $0 \leqq x_1, x_2 \leqq 1$ with DIRICHLET boundary conditions along its boundary $\partial\Omega$. These boundary conditions and the initial condition at $t = 0$ are defined by the exact solutions (3.3) and (3.4).

The initial-boundary value problems were semi-discretized on a uniform grid in the (x_1, x_2)-plane using the standard symmetric differences. As gridsize we chose $h = 1/20$ for equation (3.1) and $h = 1/20$, $h = 1/40$ for equation (3.2). This results in systems of respectively 361 and 1521 ordinary differential equations of the form (1.1). The solutions of these systems can be derived from $u(t, x_1, x_2)$ by restricting (x_1, x_2) to the grid points (this is a consequence of x_1 and x_2 occurring quadratically in (3.3) and linearly in (3.4)).

The spectral radius $\sigma(J)$ needed in the stability condition was approximated by respectively

$$\sigma = \frac{8}{h^2}, \qquad \sigma = \max_{\Gamma_h} \frac{24u^2}{h^2} \cong \frac{24}{h^2}. \tag{3.5}$$

3.2. Methods used

Three RK formulas combined with several modified right hand side functions $\boldsymbol{f}$ were applied to the problems (3.1) and (3.2). In addition, we used the standard ADI method of PEACEMAN and RACHFORD.

For order one and order two the RK formulas used were proposed in [4] and are of the form

$$\begin{aligned} &\boldsymbol{y}_{n+1}^{(0)} = \boldsymbol{y}_n, \qquad \boldsymbol{y}_{n+1}^{(1)} = \boldsymbol{y}_n + \tilde{\mu}_1 \tau \boldsymbol{f}^*(t_n, \boldsymbol{y}_n), \\ &\boldsymbol{y}_{n+1}^{(j)} = \mu_j \boldsymbol{y}_{n+1}^{(j-1)} + \nu_j \boldsymbol{y}_{n+1}^{(j-2)} + \gamma_j \boldsymbol{y}_n + \tilde{\mu}_j \tau \boldsymbol{f}^*(t_n + \theta_{j-1}\tau, \boldsymbol{y}_{n+1}^{(j-1)}) + \tilde{\gamma}_j \tau \boldsymbol{f}^*(t_n, \boldsymbol{y}_n), \\ &\theta_0 = 0, \qquad \theta_1 = \tilde{\mu}_1, \qquad \theta_j = \mu_j \theta_{j-1} + \nu_j \theta_{j-2} + \tilde{\mu}_j + \tilde{\gamma}_j, \qquad j = 2, \dots, m, \\ &\boldsymbol{y}_{n+1} = \alpha_0 \boldsymbol{y}_n + \alpha_l \boldsymbol{y}_{n+1}^{(l)} + \alpha_m \boldsymbol{y}_{n+1}^{(m)}, \qquad m = 3l. \end{aligned} \tag{3.6}$$

The **first** formula is of first order, has $\tilde{R}_m^{(1)}(z)$ as its stability polynomial (cf. (2.5)) and is defined by

$$\begin{aligned} &\tilde{\mu}_1 = \frac{w_0 + 1}{\beta w_0}, \qquad \mu_j = 2w_0 \frac{T_{j-1}(w_0)}{T_j(w_0)}, \qquad \nu_j = 1 - \mu_j, \quad \gamma_j = 0, \\ &\tilde{\mu}_j = \frac{w_0 + 1}{\beta w_0}\mu_j, \qquad \tilde{\gamma}_j = 0, \qquad j = 2, \dots, m, \end{aligned} \tag{3.7}$$

$$\alpha_0 = 0, \quad \alpha_l = 0, \quad \alpha_m = 1, \quad w_0 = 1 + \frac{1}{20m^2}, \qquad \beta \cong 1.93m^2, \qquad R_m(z) = \tilde{R}_m^{(1)}(z).$$

The second formula is second order accurate, has $\tilde{A}_m^{(2)}(z)$ as its stability polynomial (cf. (2.6)) and is defined by

$$\tilde{\mu}_1 = \frac{a_m(w_0+1)\,T_m(w_0)}{\beta w_0}, \qquad \mu_j = 2w_0 \frac{T_{j-1}(w_0)}{T_j(w_0)}, \qquad \nu_j = 1-\mu_j, \quad \gamma_j = 0,$$
$$\tilde{\mu}_j = \frac{w_0+1}{\beta w_0}\mu_j, \qquad \tilde{\gamma}_j = -a_0\tilde{\mu}_j, \quad j = 2, \dots, m, \tag{3.8}$$
$$\alpha_0 = 0, \qquad \alpha_l = 0, \qquad \alpha_m = 1, \qquad w_0 = 1 + \frac{2}{13m^2}, \qquad \beta \cong 0.65m^2, \qquad R_m(z) = \tilde{A}_m^{(2)}(z).$$

The third order consistent polynomial $\tilde{B}_m^{(3)}(z)$ is used as stability polynomial for the third method, making it third order accurate for *linear* differential equations. By making all intermediate stages second order consistent it can easily be proved that non-linear problems are also integrated with third order accuracy. This can be achieved for all stages except for the first one. In this first stage, however, the parameter $\tilde{\mu}_1$ is free, hence the third order behaviour of the scheme can be approximated as accurate as desired by choosing $\tilde{\mu}_1$ sufficiently small. A scheme which satisfies the second order conditions in all intermediate stages (except for the first one) and which is third order consistent in t_{n+1} is given by (cf. [6])

$$b_j = \frac{T_j''(w_0)}{[T_j'(w_0)]^2}, \qquad j = 2, \dots, m, \qquad b_0 = b_1 = b_2, \tag{3.9}$$
$$\tilde{\mu}_1 = b_1 w_1, \qquad \mu_j = 2w_0\frac{b_j}{b_{j-1}}, \qquad \nu_j = -\frac{b_j}{b_{j-2}}, \qquad \gamma_j = 1-\mu_j-\nu_j,$$
$$\tilde{\mu}_j = 2w_1\frac{b_j}{b_{j-1}}, \qquad \tilde{\gamma}_j = -\tilde{\mu}_j(1-b_{j-1}T_{j-1}(w_0)), \qquad j = 2, \dots, m,$$
$$\alpha_m = a_m/b_m, \qquad \alpha_l = a_l/b_l, \qquad \alpha_0 = a_0 - \alpha_l(1-b_lT_l(w_0)) - \alpha_m(1-b_mT_m(w_0)),$$
$$w_0 = 1 + \frac{2}{13m^2}, \qquad \beta \cong 0.35m^2, \qquad R_m(z) = \tilde{B}_m^{(3)}(z).$$

The methods defined by (3.7) and (3.8) are internally stable for unlimited large m-values [4]. Along the same lines the third order method defined by (3.9) can be proved to be internally stable for $m = 3l$ with l arbitrarily large.

The modified right hand side functions $\boldsymbol{f}^*$ used in the experiments are taken from Section 2.3 and specified in the tables of results given in Section 3.3.

The number of stages in the RK methods was chosen according to the stability condition (1.2), i.e.

$$m = \left[\sqrt{\frac{\tau\sigma}{c_p}} + 1\right], \qquad c_1 \cong 1.93, \quad c_2 \cong 0.65, \quad c_3 \cong 0.35, \tag{3.10}$$

where $[x]$ denotes the greatest integer less than or equal to x. Notice that the constants c_2 and c_3 are less than those mentioned in the introduction which correspond to maximally stable stability polynomials. Since these optimal polynomials are not available in closed form for $p > 1$ we used the approximations $\tilde{A}_m^{(2)}(z)$ and $\tilde{B}_m^{(3)}(z)$ at the cost of a lower c_p-value.

3.3 Numerical results

In the tables of results given belong the values of the pair $m\backslash A$ are listed where A denotes the number of correct digits in the numerical solution, i.e.

$$A = \min_{\Gamma_h} \left(-{}^{10}\log|\text{exact solution-numerical solution}|\right). \tag{3.11}$$

In the case of the ADI method m will denote the number of $\boldsymbol{f}$-evaluations per step needed in the NEWTON iteration processes. The various methods in the tables will be denoted by {integration formula; modification $\boldsymbol{f}^*$}. To compare the efficiency of the MRK and ADI methods we list for the most efficient MRK method the efficiency rate with respect to the ADI method. Actually the listed efficiency rates can be increased because in the ADI method we did not take into account the effort involved in evaluating the JACOBIan matrix and the solution of the implicit equations (see section 3.4).

In table 3.1 results are listed for the linear problem (3.1). It shows that the constant-time modification (2.12a) causes an unacceptably large drop in accuracy in spite of the fact that the generating formula (3.7) is of first order and the deviation from the classical RK formula of second order in τ (note that we choose $\theta = \beta_2$, cf. (2.16)). The modification (2.12b) in which the boundary field $\boldsymbol{g}(t, \boldsymbol{y})$ and the internal field $\boldsymbol{y}$ are tuned to the same time level, performs better but evidently the local deviation error (2.9) is still much larger than the local error of the RK formula. The second order version (2.17b) of (2.12b) turns out to produce a deviation error which is small with respect to the errors of the classical formulas (3.7), (3.8) and (3.9) (we note that the A-values of the RK solution and the MRK solution differ by at most .30 if the global deviation error is roughly equal to the global error of the generating formulas). Since in this example the modification (2.17b) with $\bar{\theta} = 0$, $\bar{\bar{\theta}} = 1$ is hardly more expensive than (2.12b) we may conclude that the *second order modification* (2.17b) *combined with the second order* RK *formula is the most*

Table 3.1. Results for equation (3.1) when discretized on a grid with $h = 1/20$

method	$\tau = 1$	$\tau = \frac{1}{12}$	$\tau = \frac{1}{35}$	$\tau = \frac{1}{70}$
$\{(3.7); f^* = f\}$	41\1.39	12\2.74	7\3.52	
$\{(3.7); (2.12\text{a}), \theta = 0\}$	41\−.06	12\.63	7\1.21	
$\{(3.7); (2.12\text{a}), \theta = \beta_2\}$	41\.07	12\.63	7\1.22	
$\{(3.7); (2.12\text{b}), \theta = \beta_2\}$	41\.55	12\2.19	7\3.26	
$\{(3.7); (2.17\text{b}), \bar{\theta} = 0, \bar{\bar{\theta}} = 1\}$	41\1.15	12\2.71	7\3.51	
$\{(3.8); f^* = f\}$	71\2.02	21\3.70	12\4.49	9\5.08
$\{(3.8); (2.12\text{b}), \theta = 1/2\}$	71\.82	21\2.01	12\2.61	9\3.27
$\{(3.8); (2.17\text{b}), \bar{\theta} = 0, \bar{\bar{\theta}} = 1\}$	71\2.29	21\3.53	12\4.43	9\5.02
$\{(3.9); (f^* = f\}$	99\1.02	30\3.92	18\5.19	12\6.20
$\{(3.9); (2.17\text{b}), (2.19')\}$	99\1.02	30\3.82	18\5.15	12\6.21
$\{\text{ADI}; f\}$	2\.79	2\2.81	2\3.74	2\4.34
Efficiency rate	0.19	0.22	0.37	0.49

efficient MRK *method*. The efficiency rate listed in table 3.1 shows that this MRK method is competitive with the ADI method if in the low accuracy range the evaluation of f^* costs 20% of the evaluation of f and about 45% in the high accuracy range.

Table 3.2 shows that the deviation error of the first order modifications is relatively small in comparison with the error of the generating formulas. It is also clearly seen that the second order RK formula (3.8) is considerably more sensitive to modification than the first order RK formula (3.7). The third order formula (3.9) is even more accurate in its modified form. The ADI method developed instabilities for $\tau \geqq 1/40$ (indicated by * in the tables of results). In the higher accuracy region the MRK method $\{(3.9), (2.17\text{b}), (2.19')\}$ is competitive with the ADI method even if the evaluation of f^* and of f requires the same computational effort. In the lower accuracy region ($A \in [1, 3]$ say) the MRK method $\{(3.8), (2.12\text{b}), \theta = 1/2\}$ is at least competitive with ADI if the evaluation of f^* is about four times "cheaper" than the evaluation of f.

Table 3.2. Results for equation (3.2) with $h = 1/20$

method	$\tau = 1$	$\tau = \frac{1}{10}$	$\tau = \frac{1}{20}$	$\tau = \frac{1}{40}$	$\tau = \frac{1}{80}$	$\tau = \frac{1}{160}$
$\{(3.7); f^* = f\}$	71\−.23	23\.87	16\1.25	12\1.56	8\1.86	
$\{(3.7); (2.12\text{b}), \theta = \beta_2\}$	71\−.16	23\.83	16\1.24	12\1.56	8\1.86	
$\{(3.8); f = f^*\}$	122\−.26	38\1.41	28\2.05	20\2.89	14\3.66	10\4.26
$\{(3.8); (2.12\text{b}), \theta = 1/2\}$	122\−.13	38\.78	28\1.51	20\2.39	14\3.45	10\4.23
$\{(3.9); f^* = f\}$	168\−.32	54\.67	39\2.03	27\4.07	21\5.07	15\5.94
$\{(3.9); (2.17\text{b}); (2.19')\}$	168\−2.38	54\.65	39\2.05	27\4.25	21\5.30	15\6.21
$\{\text{ADI}; f\}$	10*	10*	10*	10*	2\2.11	2\2.73
					4\3.18	4\4.25
Efficiency rate				0.25	0.38	0.38

Table 3.3. Results for equation (3.2) with $h = 1/40$

method	$\tau = 1$	$\tau = \frac{1}{10}$	$\tau = \frac{1}{20}$	$\tau = \frac{1}{40}$	$\tau = \frac{1}{80}$	$\tau = \frac{1}{160}$
$\{(3.7); f^* = f\}$	142\−.21	45\.85	32\1.24	23\1.56	16\1.86	12\2.16
$\{(3.7); (2.12\text{b}), \theta = \beta_2\}$	142\−.15	45\.81	32\1.23	23\1.56	16\1.86	12\2.16
$\{(3.8); f^* = f\}$	244\−.31	77\1.36	55\2.00	39\2.83	28\3.67	20\4.28
$\{(3.8); (2.12\text{b}), \theta = 1/2\}$	244\−.14	77\.71	55\1.50	39\2.36	28\3.27	20\4.24
$\{\text{ADI}; f\}$	10*	10*	10*	10*	4\2.67	4\3.71
Efficiency rate					0.23	0.37

In table 3.3 the experiments listed in table 3.2 are repeated for $h = 1/40$. We see that for larger values of $\tau\sigma$ the accuracy of the RK and MRK methods is nearly preserved whereas the ADI method shows a substantial decrease in accuracy.

3.4. Concluding remarks

The results presented in the preceding section suggest that stabilized RK methods could be an alternative for ADI type methods if the problem at hand admits a time saving modification of the right hand side function. Such a statement asks for a more detailed comparison of the computational aspects of the two classes of methods.

Computational effort per step: For r-dimensional problems ($r \geqq 2$), an ADI method requires per integration step at least f-evaluations but usually $2r$ or more evaluations if the problem is nonlinear. In addition, the evaluation of the Jacobian matrices needed in the Newton iteration processes requires $3r + 1$ f-evaluations. Of course, the Jacobian matrices do not need updating in each integration step but even for moderate nonlinear problems, particularly when the space discretization parameter h is small, a decrease in the rate of convergence of the Newton process may be induced which in turn leads to extra f-evaluations. Finally, each time the JACOBIANS are updated decompositions should be made in order to solve the one-dimensional implicit relations. Again the computational effort involved is larger as r is larger.

The MRK method requires m f^*-evaluations per integration step where m is given by (3.10). Because σ is proportional to r (for the model problem $u_t = \Delta u$ we have $\sigma = 4r/h^2$), we roughly find for parabolic problems

$$m \sim \sqrt{\tau\sigma} = O\left(\sqrt{\frac{\tau r}{h^2}}\right) \quad \text{as} \quad h \to 0 . \tag{3.10'}$$

In addition, we need an approximation to the spectral radius of the JACOBIAN which will require additional f^*-evaluations (for instance, the power method might be used). Just as in the case of ADI methods where the JACOBIANS do not need updating in each step, the MRK method only occasionally requires updating of σ. However, one should be careful not to introduce instabilities by a too careless update strategy. Therefore it is recommended to use a slightly overestimated value of σ. In fact, an overestimation of σ by $s\%$ leads to an increase of $s/2\%$ for the value of m.

Accuracy: We consider the accuracy of the ADI and MRK methods in terms of the number N of f-evaluations and the number N^* of f^*-evaluations needed to integrate the interval [0, 1]. From the discussion above it follows that

$$N = \nu r \frac{1}{\tau}, \qquad N^* \cong \sqrt{\tau\sigma}\,\frac{1}{\tau}, \tag{3.12}$$

where ν denotes the number of NEWTON iterations. Let $u(t_n)$ denote the solution of the partial differential equation at the grid points for $t = t_n$. Then the global errors of the ADI and MRK methods are respectively given by

$$\varepsilon_n \cong u(t_n) - y(t_n) + \delta_n(h) + c_n(h)\,\tau^2 = u(t_n) - y(t_n) + \delta_n(h) + \left(\frac{\nu r}{N}\right)^2 c_n(h) \tag{3.13}$$

and

$$\varepsilon_n^* \cong u(t_n) - y(t_n) + c_n^*(h)\,\tau^p = u(t_n) - y(t_n) + \left(\frac{\sigma}{(N^*)^2}\right)^p c_n^*(h) \tag{3.14}$$

where $\delta_n(h)$ is the iteration error of the NEWTON process left after ν iterations, p is the order of the MRK method and $c_n(h)$, $c_n^*(h)$ represent the principal error constants of the ADI and MRK time integrators (these quantities will depend on h because f and f^* depend on h).

Let h be fixed and let $\tau \to 0$ (N, $N^* \to \infty$). From the expressions above we see that the global error of the higher order MRK methods ($p \geqq 2$) converges much faster to the "space error" $u(t_n) - y(t_n)$ than the global error of the ADI method does. Hence, in problems where the space error is small (i.e. on fine grids with h small) one shall wish small time errors, that is large N or N^* values, and in such problems the higher order MRK methods will eventually become superior even when the f^*-evaluations are only slightly less cheaper than the f-evaluations.

The expressions (3.13) and (3.14) also clearly show the effect of nonlinearities (increase of ν) and the number of dimensions on the accuracy of the ADI method, whereas the MRK method is not affected at all.

It is difficult to make general statements on the error constants c_n and c_n^*. But in view of the work of FAIRWEATHER and MITCHELL [1] it is known that splitting methods lose accuracy when the boundary conditions are time-dependent and if $h \to 0$. This phenomenon can be interpreted as an increase of $||c_n(h)||$ as $h \to 0$. The MRK methods with an appropriate f^* function do not have this disadvantage (see section 2.5).

Storage requirements: To implement the ADI method we need $3r + 4$ arrays of length corresponding to the number of gridpoints. We use $3r$ arrays to hold the (decomposed) JACOBIAN matrices and 4 arrays for the approximate solutions or derivatives. For the MRK methods only the size of the arrays depends on r whereas the number of arrays is independent of r. For the first, second and third order method we need 3, 4 and 6 arrays, respectively.

The final conclusion (based on our experiments) is that

(i) the second order MRK method is the most efficient one of the several MRK methods;

(ii) the MRK method is advantageous in problems where the modification results in a substantial reduction of computing time when compared with the generating RK method. We found reduction to 25% adequate in order to be competitive with the ADI method for the h-values used;

(iii) the stability behaviour of the MRK method is insensitive to large $\tau\sigma$-values (h small and/or τ large); hence low accuracy results can be obtained for large integration steps on fine grids and in strongly non-linear problems, while the ADI method behaves unstable in this situation.

References

1 FAIRWEATHER, G.; MITCHELL, A. R., A new computational procedure for ADI methods, SIAM J. Numer. Analysis 4, 163–170 (1967).

2 HOUWEN, P. J. VAN DER, Stabilized RUNGE-KUTTA methods for second order differential equations without first derivatives, SIAM J. Numer. Analysis 16, 523–537 (1979).

3 HOUWEN, P. J. VAN DER, Explicit RUNGE-KUTTA formulas with increased stability boundaries, Numer. Math. 20, 149–164 (1972).

4 HOUWEN, P. J. VAN DER; SOMMEIJER, B. P., On the internal stability of explicit, m-stage RUNGE-KUTTA methods for large values of m, ZAMM 60, 10; 479–485 (1980).

5 SHAMPINE, L. F., Storage reduction for RUNGE-KUTTA codes, ACM Trans. Math. Software 5, 245–250 (1979).

6 SOMMEIJER, B. P.; VERWER, J. G., A performance evaluation of a class of Runge-Kutta-Chebyshev methods for solving semi-discrete parabolic differential equations, Report NW 91/20 Mathematisch Centrum, Amsterdam (1980)

Eingereicht: 17. 12. 1979, korrigierte Fassung 31. 7. 1980

Anschriften: Professor Dr. P. J. VAN DER HOUWEN, Dr. B. P. SOMMEIJER, Stichting Mathematisch Centrum, Kruislaan 413, 1098 SJ Amsterdam, Niederlande

KLEINE MITTEILUNGEN

ZAMM 61, 115–117 (1981)

W. Mönch

Anwendung des Schulz-Verfahrens zur Nachkorrektur einer näherungsweise berechneten verallgemeinerten Inversen einer Matrix

Das auf Schulz [9] zurückgehende Verfahren

$$\text{(VO)} \quad X_{k+1} = X_k + X_k(I - AX_k), \qquad k = 0, 1, \dots,$$

zur iterativen Berechnung der Inversen A^{-1} einer quadratischen, regulären Matrix A wird in der Praxis hauptsächlich zur Nachkorrektur einer (etwa mit einem direkten Verfahren berechneten) Näherung Z für A^{-1} eingesetzt (Startwert für (VO): $X_0 = Z$). Auf Grund seiner quadratischen Konvergenz ist es den linear konvergenten Verfahren zur Nachkorrektur von Näherungslösungen linearer Gleichungssysteme überlegen (s. Forsythe, Moler [6]). Steht keine hinreichend gute Näherung Z für A^{-1} zur Verfügung, so ist das Verfahren (VO) — falls es überhaupt konvergiert — i. a. wesentlich aufwendiger als direkte Verfahren zur Berechnung von A^{-1}, gekoppelt mit einer iterativen Nachkorrektur.

(VO) kann auch zur Berechnung einer verallgemeinerten Inversen einer beliebigen Matrix $A \in R^{m\times n}$ herangezogen werden (s. Ben-Israel [1, 2], Ben-Israel, Cohen [3]). Bisherige numerische Tests und Stabilitätsuntersuchungen (s. Söderström, Stewart [11], Shinozaki, Sibuya, Tanabe [10]) zeigen jedoch, daß das Verfahren — insbesondere bei Matrizen mit Rangabfall — nicht geeignet ist. Allerdings wurde dabei stets von einer Startmatrix $X_0 = \alpha A^T$ ausgegangen, die zwar bei geeignetem Parameter α *theoretisch* Konvergenz sichert, aber i. a. keine gute Näherung für eine verallgemeinerte Inverse von A ist.

Hier soll gezeigt werden, wie man das Verfahren (VO) entsprechend dem Vorgehen von Zielke [12, 13] für die Nachkorrektur einer mit einem direkten Verfahren berechneten Näherung Z für eine verallgemeinerte Inverse von A einsetzen kann.

Es bezeichne $R^{m\times n}$ die Menge aller Matrizen vom Typ (m, n) mit reellen Elementen, Rang (A) den Rang einer Matrix A, A^T die zu A transponierte Matrix, cond $(A) = ||A||_2\,||A^+||_2$ die Spektralkonditionszahl von A und I die Einheitsmatrix (passenden Typs). Vorgegeben seien eine Matrix $A \in R^{m\times n}$ mit $r =$ Rang $(A) \leqq \min(m, n)$ sowie eine beliebige Näherung $Z \in R^{n\times m}$ für eine verallgemeinerte Inverse von A. Unter einer *verallgemeinerten Inversen von* A wird hier eine Lösung $X \in R^{n\times m}$ einiger der (bzw. aller) vier Penrose-Gleichungen

$$AXA = A, \qquad XAX = X, \tag{1), (2}$$

$$(AX)^T = AX, \qquad (XA)^T = XA, \tag{3), (4}$$

(s. z. B. Ben-Israel, Greville [4]) verstanden. Bezeichnet wird z. B. eine Lösung von (1), (2) und (3), kurz (1, 2, 3)-Inverse von A genannt, mit $A^{1,2,3}$; für die Bezeichnung der Moore-Penrose *Pseudoinversen* $A^{1,2,3,4}$ wird insbesondere A^+ verwendet.

Ausgehend von einer Näherung $Z \in R^{n\times m}$ für eine verallgemeinerte Inverse von A wird hier X_0 zunächst so gewählt, daß die Forderungen

(F1) X_0 ist ebenfalls eine Näherung für eine verallgemeinerte Inverse von A;

(F2) die Spalten von X_0^T liegen im Spaltenraum von A und

(F3) (VO) konvergiert von X_0 aus gegen eine verallgemeinerte Inverse von A,

erfüllt sind. Es wird

$$X_0 = \alpha Z(AZ)^T \quad \text{mit} \quad \alpha \in R^1 \tag{5}$$

gesetzt.

Die erste Forderung an X_0 ist damit in folgendem Sinne erfüllt:

Aussage 1: *Für $X_0 = \alpha Z(AZ)^T$ gelten die Aussagen:*

(i) *Falls $Z \in R^{n\times m}$ beliebig, dann $X_0 = A^3$.*

(ii) *Falls $Z = A^{1,3}$, dann $X_0 = \alpha A^{1,3}$.*

(iii) *Falls $Z = A^{2,3}$, dann $X_0 = \alpha A^{2,3}$.*

(iv) *Falls $Z = A^{1,2,3}$, dann $X_0 = \alpha A^{1,2,3}$.*

(v) *Falls $Z = A^+$, dann $X_0 = \alpha A^+$.*

Die Richtigkeit dieser Aussagen bestätigt man unmittelbar durch Einsetzen in die entsprechenden Definitionsgleichungen (1) bis (4). Die Forderung (F2) ist stets erfüllt.

Zum Nachweis der Konvergenz von (VO) wird eine *Singulärwertzerlegung* von AZ benötigt:

$$\begin{aligned} &AZ = USV^T \quad \text{mit} \quad U, V \in R^{m\times m} \quad \text{orthogonal},\\ &S = \operatorname{diag}(s_1, \dots, s_m) \in R^{m\times m},\\ &s_1 \geqq \dots \geqq s_r > s_{r+1} = \dots = s_m = 0, \qquad r = \text{Rang}(AZ). \end{aligned} \tag{6}$$

Bezüglich der Konvergenz von (VO) mit dem Startwert (5) gilt dann der

Satz 2: *Es seien $A \in R^{m\times n}$ und $Z \in R^{n\times m}$ vorgegeben, und es wird $X_0 = \alpha Z(AZ)^T$ gesetzt. Dann konvergiert die nach (VO) berechnete Folge $\{X_k\}$ genau dann mit der Konvergenzordnung $\tau = 2$ gegen $Z(AZ)^+$, wenn $0 < \alpha < 2/s_1^2$ ist.*

Beweis: Aus der Beziehung

$$X_k = Z(AZ)^+ - Z(AZ)^+\,[(AZ)\,(AZ)^+ - AX_0]^{2^k}, \quad k = 0, 1, \dots, \tag{7}$$

deren Richtigkeit man unter Beachtung von

$$X_k = Z(AZ)^+\,AX_k \quad \text{und} \quad X_k = X_k(AZ)\,(AZ)^+, \quad k = 0, 1, \dots, \tag{8}$$

durch vollständige Induktion bestätigt, folgt mit der Singulärwertzerlegung (6) von (AZ):

$$X_k = Z(AZ)^+ - ZVS^+[I_r - \alpha SS^T]^{2^k}\,U^T.$$

Ist nun $0 < \alpha < 2/s_1^2$, dann gilt $|1 - \alpha s_i^2| < 1$ $(i = 1, \dots, r)$ und damit $\lim\,[I_r - \alpha SS^T]^{2^k} = 0$, so daß in diesem Fall $\{X_k\}$ gegen $Z(AZ)^+$ konvergiert.

Gilt andererseits $\lim X_k = Z(AZ)^+$, dann folgt wegen

$$\begin{aligned} \lim\,[I_r - \alpha SS^T]^{2^k} &= \lim U^T(AZ)\,(AZ)^+\,[(AZ)\,(AZ)^+ - AX_0]^{2^k}\,U\\ &= \lim U^TA[Z(AZ)^+ - X_k]\,U = 0, \end{aligned}$$

daß $|1 - \alpha s_i^2| < 1$ $(i = 1, \dots, r)$, also $0 < \alpha < 2/s_1^2$ erfüllt sein muß.

Die Konvergenzordnung $\tau = 2$ liest man unmittelbar aus (7) ab.

Es soll nun angegeben werden, welchen der vier Penrose-Gleichungen der Grenzwert $\lim X_k = Z(AZ)^+$ — falls er existiert — genügt.

Aussage 3: *Für $X = Z(AZ)^+$ gelten die folgenden Aussagen:*

(i) *Falls $Z \in R^{n\times m}$ beliebig, dann $X = A^{2,3}$.*

(ii) *Falls Rang $(Z) = n$, dann $X = A^{1,2,3}$.*

(iii) *Falls Rang $(AZ) = m$, dann $X = A^{1,2,3}$.*

(iv) *Falls Rang $(A) =$ Rang $(Z) = n$, dann $X = A^+$.*

Beweis:

(i): Es ist $XAX = Z(AZ)^+\,AZ(AZ)^+ = Z(AZ)^+ = X$ und

$$(AX)^T = [AZ(AZ)^+]^T = AZ(AZ)^+ = AX, \quad \text{d. h.} \quad X = A^{2,3}.$$

(ii): Da Rang $(Z) = n$ ist, existiert $(ZZ^T)^{-1}$. Damit ist $AXA = AZ(AZ)^+A(ZZ^T)\,(ZZ^T)^{-1} = (AZ)\,(AZ)^+\,(AZ)\,Z^T(ZZ^T)^{-1} = AZZ^T(ZZ^T)^{-1} = A$, also $X = A^1$. Mit (i) ergibt das insgesamt $X = A^{1,2,3}$.

(iii): Wegen Rang $(AZ) = m$ ist $X = Z(AZ)^{-1}$ und damit $AXA = AZ(AZ)^{-1}A = A$. Zusammen mit (i) folgt also $X = A^{1,2,3}$.

(iv): $X = Z(AZ)^+ = ZZ^+A^+ = A^+$, da $ZZ^+ = I$.

Bemerkungen:

1. Für das Residuum $R_0 = (AZ)\,(AZ)^+ - AX_0$ gilt $||R_0||_2 = \max\limits_{1\leqq i\leqq r} |1 - \alpha s_i^2|$. $||R_0||_2$ wird minimal für $\alpha = 2/(s_1^2 + s_r^2)$; es ist dann

$$\begin{aligned} ||R_0||_2 &= (s_1^2 - s_r^2)/(s_1^2 + s_r^2) =\\ &= [\text{cond}\,(AZ) - 1]/[\text{cond}\,(AZ) + 1] < 1. \end{aligned}$$

Falls daher cond (AZ) hinreichend klein ist, kann man — wie (7) zeigt — hinreichend schnelle Konvergenz erwarten.

2. Da mit jeder beliebigen Matrixnorm $s_1^2 \leqq ||AZ(AZ)^T||$ gilt, kann für die praktische Rechnung ohne großen Zusatzaufwand ein Parameter α bestimmt werden, der Konvergenz sichert.

3. Bei beliebiger Anfangsnäherung $Z \in R^{n\times m}$ genügt $X = Z(AZ)^+$ i. a. nur den Gleichungen (2) und (3). Durch eine geeignete Regularisierung von Z kann man jedoch stets erreichen, daß der Grenzwert X eine (1, 2, 3)-Inverse von A ist. Wird ZZ^T durch $(\lambda I + ZZ^T)$ mit einem Regularisierungsparameter $\lambda > 0$ ersetzt, dann ist $X_0 = \lambda(\alpha A^T) + \alpha Z(AZ)^T$, d. h. X_0 ist aus der von BEN-ISRAEL [1, 2] und der hier verwendeten Startmatrix zusammengesetzt.

4. Eine Übertragung der Betrachtungen über (VO) auf Verfahren mit einer Konvergenzordnung $\tau > 2$ (s. z. B. BEN-ISRAEL, GREVILLE [4]) ist möglich, allerdings im Zusammenhang mit der Nachkorrektur nicht so von Interesse.

Für die FROBENIUS-*Norm des Defektes* $(I - AX_k)$ gilt wegen

$$I - AX_k = [I - AX_0]^{2^k} = [I - \alpha AZ(AZ)^T]^{2^k} = U[I - \alpha SS^T]^{2^k} U^T$$

die Beziehung

$$\|I - AX_k\|_F^2 = \sum_{i=1}^{r} (1 - \alpha s_i^2)^{2^k} + (m - r)\,. \tag{9}$$

Daraus liest man unmittelbar die folgende Aussage ab, die z. B. für einen Abbruchtest genutzt werden kann.

Aussage 4: *Falls* $X_0 = \alpha Z(AZ)^T$ *mit* $0 < \alpha < 2/s_1^2$ *gewählt wird, gilt für die nach* (VO) *berechnete Folge* $\{X_k\}$

$$m - r \leqq \|I - AX_{k+1}\|_F^2 \leqq \leqq \|I - AX_k\|_F^2 \leqq \ldots \leqq \|I - AX_0\|_F^2 < m\,.$$

Aussage 1 legt die Wahl $\alpha = 1$ nahe (es entfällt dann eine Abschätzung von s_1). Die folgende, sich aus (10) ergebende Aussage gibt in gewissen Fällen an, ob X_0 mit $\alpha = 1$ im Konvergenzbereich von (VO) oder außerhalb davon liegt.

Aussage 5: *Es sei* $X_0 = \alpha Z(AZ)^T$.

(i) *Falls* $\|I - AX_0\|_F^2 \geqq m$, *dann ist* $\alpha \geqq 2/s_1^2$.

(ii) *Falls* $\|I - AX_0\|_F^2 < m - r + 1$, *dann ist* $0 < \alpha < 2/s_1^2$.

Wird statt (5) als Startmatrix

$$X_0 = \alpha(ZA)^T Z, \qquad \alpha \in R^1 \tag{10}$$

gewählt, dann ist wegen $X_0^T = \alpha Z^T(A^T Z^T)^T$ eine Zurückführung auf den bisher betrachteten Fall möglich, wenn man nur A durch A^T und Z durch Z^T ersetzt. Im Falle der Konvergenz von (VO) mit (10) gilt $\lim X_k = (ZA)^+ Z = A^{2,4}$ bei beliebigem $Z \in R^{n\times m}$.

Ist insbesondere die MOORE-PENROSE-Pseudoinverse A^+ gesucht, so kann man bei beliebig vorgegebener Näherung $Z \in R^{n\times m}$ (mit Rang $(Z) = n$, vgl. Bemerkung 3.) zunächst nach (VO) mit dem Startwert $X_0 = \alpha Z(AZ)^T$ die (1,2,3)-Inverse $X = Z(AZ)^+$ von A zumindest näherungsweise berechnen und anschließend (VO) mit $\tilde{X}_0 = \tilde{\alpha}(XA)^T X$ (vgl. (10)) neu starten. Falls $0 < \tilde{\alpha} < 2/\tilde{s}_1^2$ ($\tilde{s}_1$ größter singulärer Wert von XA) ist, konvergiert die nach dem Neustart von (VO) berechnete Folge $\{\tilde{X}_k\}$ gegen $Y = (XA)^+X$. Wie die folgende Aussage zeigt, ist $Y = A^+$.

Aussage 6: *Es sei* $X \in R^{n\times m}$ *eine* (1, 2, 3)-*Inverse von* A. *Dann gilt* $A^+ = (XA)^+ X$.

Beweis: Die Matrix $Y := (XA)^+ X$ genügt den vier Definitionsgleichungen (1) bis (4) für A^+:

(i): $AYA = (AXA)YA = A(XA)(XA)^+(XA) = AXA = A.$

(ii): $YAY = (XA)^+XA(XA)^+X = (XA)^+X = Y.$

(iii): $AY = A(XA)^+X = A(XA)^+(XAX) = A(XA)^+X(AX) = AY(AX) = (AYA)X = AX = (AX)^T = (AY)^T.$

(iv) $YA = (XA)^+XA = [(XA)^+XA]^T = (YA)^T.$

Ist speziell die MOORE-PENROSE-Pseudoinverse A^+ von A gesucht, und wird A^+ durch Lösen linearer Ausgleichsprobleme $AX = I$ im Sinne kleinster Fehlerquadrate bestimmt, so liefern die dafür bekannten, linear konvergenten Verfahren für eine Nachkorrektur von Näherungslösungen (s. z. B. [8] und [5]) im allgemeinen nur eine (1,2,3)-Inverse von A. Das SCHULZ-Verfahren (VO) bietet daher neben der höheren Konvergenzordnung noch den Vorteil, daß die damit in der oben beschriebenen Weise berechneten Näherungen gegen A^+ konvergieren.

Wird (VO) in dem hier betrachteten Sinne zur Nachkorrektur eingesetzt, so ist die Iteration i. a. nur dann erfolgreich, wenn bestimmte Operationen mit höherer Genauigkeit ausgeführt werden. Eine erste Möglichkeit besteht in der Berechnung von $\delta X_k = X_k(I - AX_k)$ mit höherer Genauigkeit. Im Falle der Konvergenz gilt bei beliebigen Matrizen A und Z für den Grenzwert von $\{\delta X_k\}$: $\lim \delta X_k = 0$. Da

$$\lim (I - AX_k) = I - (AZ)(AZ)^+ \neq 0$$

für Rang $(AZ) < m$ ist, genügt es i. a. nicht, nur $(I - AX_k)$ mit höherer Genauigkeit zu berechnen. Eine zweite Variante, bei deren Implementierung nur die Möglichkeit einer Skalarproduktberechnung mit höherer Genauigkeit vorhanden sein muß, entsteht bei Übertragung der für lineare Ausgleichsprobleme bekannten Algorithmen (s. GOLUB, SAUNDERS [8], BJÖRCK [5]).

(VO) wird äquivalent (im Sinne der Analysis) umgeformt in das Verfahren

(V1) *Startwerte*: $X_0 = \alpha Z(AZ)^T \in R^{n\times m}$ (bzw. $X_0 = \alpha(ZA)^T Z$) $D_0 \in R^{m\times m}$ beliebig

Iteration: $X_{k+1} = X_k + X_k(I - AX_k - D_k) + X_kD_k,$
$D_{k+1} = D_k + (I - AX_k - D_k) - A[X_k(I - AX_k - D_k) + X_kD_k], \quad k = 0, 1, \ldots,$

wobei $(I - AX_k - D_k)$ und x_kD_k mit höherer Genauigkeit und alle übrigen Terme mit einfacher Genauigkeit berechnet werden.

Falls $\lim X_k = Z(AZ)^+ = X$ existiert, konvergiert die Folge $\{D_k\}$ (bei beliebigem $A \in R^{m\times n}$, $Z \in R^{n\times m}$ und $D_0 \in R^{m\times m}$) gegen $I - AX = I - AZ(AZ)^+$. Das folgt unmittelbar aus

$$D_{k+1} - [I - (AZ)(AZ)^+] = [(AZ)(AZ)^+ - AX_k]^2 \quad (k = 0, 1, \ldots)\,.$$

Daher sind $D_0 = I - AX_0$ (mit einfacher Genauigkeit) und $D_0 = 0$ (d. h. $D_1 = I - AX_1$) als Startmatrizen D_0 naheliegend.

Numerische Tests: Es wurde die von ZIELKE [12] angegebene Testmatrix

$$A = \begin{pmatrix} a & a & a-1 & a \\ a+1 & a & a & a \\ a & a & a-1 & a \\ a+1 & a & a & a \\ a+1 & a+1 & a & a+1 \end{pmatrix}$$

mit verschiedenen reellen Parametern a benutzt; Rang $(A) = 3$, $\text{cond}_F(A) = \|A\|_F \|A^+\|_F \approx 7.75\,|a|^2$ für $|a| \gg 1$.

Jede Matrix der Form

$$A^{1,2,3} = \frac{1}{4}\begin{pmatrix} 2a & 2 & 2a & 2 & -4a \\ \gamma & -1+\delta & \gamma & -1+\delta & 2+\mu \\ -2a-2 & 0 & -2a-2 & 0 & 4a \\ -\gamma & -1-\delta & -\gamma & -1-\delta & 2-\mu \end{pmatrix}$$

mit beliebigen reellen Parametern γ, δ und μ ist eine (1, 2, 3)-Inverse von A mit $\|I - AA^{1,2,3}\|_F^2 = 2$; für $\gamma = \delta = \mu = 0$ erhält man insbesondere A^+.

Die Anfangsnäherung Z für eine verallgemeinerte Inverse von A wurde hier mittels HOUSEHOLDER-Transformationen über eine QR-Zerlegung von A als Näherung für A^+ berechnet und anschließend mit (V1), $X_0 = \alpha Z(AZ)^T$ und $D_0 = I - AX_0$ verbessert. Mindestens aller fünf Schritte wurde ein *Neustart* der Iteration vorgenommen, um damit das in [11] beschriebene Anwachsen der Rundungsfehler in Grenzen zu halten.

Von der so ermittelten Näherung X für die (1, 2, 3)-Inverse $Z(AZ)^+$ aus wurde schließlich (V1) mit $\tilde{X}_0 = \tilde{\alpha}(XA)^TX$ und $D_0 = I - A\tilde{X}_0$ neu gestartet und damit eine verbesserte Näherung Y für A^+ berechnet.

Zur Einschätzung der Genauigkeit der Ausgangsnäherung Z und der nach iterativer Verbesserung erhaltenen Näherung Y wurden die Größen $\varkappa(Z, A^+)$ und $\varkappa(Y, A^+)$ berechnet, wobei

$$\varkappa(Z, B) = -\lg (\max_{ij} q_{ij}) \qquad \text{mit}$$

$$q_{ij} = \begin{cases} |z_{ij} - b_{ij}|/|b_{ij}|, & \text{falls } b_{ij} \neq 0 \\ |z_{ij} - b_{ij}|, & \text{falls } b_{ij} = 0 \end{cases}$$

für $Z = (z_{ij})$, $B = (b_{ij}) \in R^{n\times m}$.

Die Näherung X wurde analog dazu mit der jeweils am nächsten gelegenen (1, 2, 3)-Inverse von A, für die $\varkappa(X, A^{1,2,3})$ am größten ausfällt, verglichen.

In der folgenden Tabelle sind die Ergebnisse für $a = 1$, 10, 100, 500 und 1000 zusammengefaßt. (ANZ gibt die Anzahl an Iterationsschritten an.)

Herrn Doz. Dr. G. Zielke danke ich für eine Reihe von Anregungen und Hinweisen

Tabelle

a	cond(A)	$\varkappa(Z,A^+)$	$\varkappa(X,A^{1,2,3})$	$\varkappa(Y,A^+)$	ANZ
1	.14 # 2	6.15	8.00	8.00	7
10	.80 # 3	4.68	8.00	8.00	7
100	.77 # 5	2.70	8.00	8.00	9
500	.19 # 7	1.06	8.00	8.00	12
1000	.77 # 7		8.00	8.00	15

(Maschinengenauigkeit: Gleitkommaarithmetik, 8 dezimale Mantissenstellen.)

Literatur

1 BEN-ISRAEL, A., An iterative method for computing the generalized inverse of an arbitrary matrix, Math. Comput. 19, 452–455 (1965).
2 BEN-ISRAEL, A., A note on an iterative method for generalized inversion of matrices, Math. Comput. 20, 439–440 (1966).
3 BEN-ISRAEL, A.; COHEN, D., On iterative computation of generalized inverses and associated projections, SIAM J. Numer. Analysis 3, 410–419 (1966).
4 BEN-ISRAEL, A.; GREVILLE, T. N. E., Generalized Inverses. Theory and Applications, John Wiley & Sons 1974.
5 BJÖRCK, A., Comment on the iterative refinement of least-squares solutions, J. Amer. Statist. Assoc. 73, 161–166 (1978).
6 FORSYTHE, G. E.; MOLER, C. B., Computer Solution of Linear Algebraic Systems, Prentice-Hall, Inc., Englewood Cliffs, New Jersey 1967.
7 GARNETT III, J. M.; BEN-ISRAEL, A.; YAU, S. S., A hyperpower method for computing matrix products involving the generalized inverse. SIAM J. Numer. Analysis 8, 104–109 (1971).
8 GOLUB, G. H.; SAUNDERS, M. A., Linear least squares and quadratic programming, in Integer and Nonlinear Programming, ed. JEAN ABADIE, Amsterdam-London, North-Holland Publishing Co. 229–256, 1970.
9 SCHULZ, G., Iterative Berechnung der reziproken Matrix, ZAMM 13, 57–59 (1933).
10 SHINOZAKI, N.; SIBUYA, M.; TANABE, K., Numerical algorithms for the MOORE-PENROSE inverse of a matrix: iterative methods, Ann. Inst. Statist. Math. 24, 621–629 (1972).
11 SÖDERSTRÖM, T.; STEWART, G. W., On the numerical properties of an iterative method for computing the MOORE-PENROSE generalized inverse, SIAM J. Numer. Analysis 11, 61–74 (1974).
12 ZIELKE, G., Beiträge zur Theorie und Berechnung von verallgemeinerten inversen Matrizen, Dissertation B, Halle 1977.
13 ZIELKE, G., Iterative refinement of generalized matrix inverses now practicable, SIGNUM News Letter 13, 9–10 (1978).

Eingereicht: 31. 10. 1979

Anschrift: Dr. WOLFGANG MÖNCH, Bergakademie Freiberg, Sektion Mathematik, DDR-9200 Freiberg, Bernhard-von-Cotta-Str. 2

ZAMM 61, 117–119 (1981)

M. S. PETKOVIĆ

On an Interval Newton's Method Derived from Exponential Curve Fitting

A modification of MOORE's interval method derived by exponential interpolation at three points is given. The proposed method avoids the division by the zero-interval and has convergence order almost three.

Let X_0 be an interval containing a real simple zero ξ of a function f. In [1] R. E. MOORE formulated an interval extension of NEWTON's method with

$$X_{n+1} = N(X_n) \cap X_n\,, \qquad n = 0, 1, \dots,$$

where

$$N(X) = x - \frac{f(x)}{F'(X)}\,, \qquad x \in X\,. \tag{1}$$

Usually, x is the midpoint of X. The author proved that in the neighbourhood of the root ξ this method converges quadratically to the root ξ.

$F'(X)$ is an *interval extension* of the function f', i.e. $F'(X) \supset \{f'(x) \mid x \in X\}$. Therefore $F'(X)$ can contain the zero even if f' has constant sign on X. This possibility represents a serious disadvantage of MOORE's method. Using exponential interpolation at three points, a modification of MOORE's method, which avoids this difficulty, is derived in this paper. Besides, the proposed procedure accelerates the convergence. The convergence order of the suggested method is almost three.

Assume that f is a monotonical function on the interval $X = [a, b]$ and has a real simple zero ξ in X. If the interval X is sufficiently narrow, then the function f can be efficiently interpolated by the exponential function

$$q(x) = A + B\,e^{kx} \tag{2}$$

at the points a, $c = \dfrac{a+b}{2}$ and b.

From the system of equations

$$A + B\,e^{ka} = f(a)\,,$$
$$A + B\,e^{kc} = f(c)\,,$$
$$A + B\,e^{kb} = f(b)\,,$$

we find

$$A = \frac{f(c)^2 - f(a)\,f(b)}{2f(c) - f(a) - f(b)}\,, \quad k = \frac{1}{r}\log\frac{f(b) - f(c)}{f(c) - f(a)}\,,$$

$$B = (f(a) - A)\,e^{-ka}\,,$$

where $r = \dfrac{b-a}{2}$.

C. J. F. RIDDERS [2] has used the interpolation function (2) for finding an approximate value α of the zero ξ. From the equation $q(x) = 0$ we find

$$\alpha = \frac{1}{k}\log\left(-\frac{A}{B}\right) \in [a, b]\,.$$

The proposed method is based on the substitution of x by α and of $F'(X)$ by the interval extension $Q'(X)$ of the function $x \mapsto q'(x)$ in (1).

Let $e_1 = \min(e^{ka}, e^{kb})$ and $e_2 = \max(e^{ka}, e^{kb})$. The function $x \mapsto q'(x) = Bk\,e^{kx}$ is monotonically for any x so that the interval extension Q' can be represented by

$$Q'(X) = \{q'(x) \mid x \in X\} = \left[\min_{x\in X} q'(x), \max_{x\in X} q'(x)\right]$$
$$= Bk[e_1, e_2] = [E_1, E_2]\,.$$

Since $e_1 > 0$ and $e_2 > 0$, it follows that $0 \notin Q'(X)$ for any X.

Direct substitution $Q'(X)$ instead of $F'(X)$ in (1) can cause that the interval $\alpha - f(\alpha)/Q'(X)$ does not contain the zero of the function f because of the possibility $Q'(X) \subset F'(X)$. For this reason, we shall take an extended interval $Q'(X) + I_\varepsilon$ instead of $F'(X)$. I_ε is one of the intervals $[-\varepsilon, \varepsilon]$, $[-\varepsilon, 0]$, $[0, \varepsilon]$, $[0, 0]$ $(\varepsilon > 0)$, which have to be chosen so that the interval

$$N_\varepsilon(X) = \alpha - \frac{f(\alpha)}{[E_1, E_2] + I_\varepsilon} \tag{3}$$

contains the zero ξ of f, and that $0 \notin [E_1, E_2] + I_\varepsilon$.

Let X_0 be an initial interval containing the zero ξ. On the basis of (3), we can formulate the following iterative interval method

$$X_{n+1} = N_\varepsilon(X_n) \cap X_n\,, \qquad n = 0, 1, \dots\,. \tag{4}$$

Denote the *convergence order of* RIDDERS' *method* by λ. The convergence of the interval method (4) is considered in the following theorem.

Theorem: *Let the f be a monotonical function on the interval X_0, and let X_0 contain a real simple zero ξ of f. Then the convergence order of the interval method* (4) *is $\lambda + 1$, i.e. there exists a positive constant C such that*

$$d(X_{n+1}) \leqq C d(X_n)^{\lambda+1}\,.$$

Proof: Let X be an interval containing the zero ξ. It is easy to conclude that there are some positive numbers p_1 and p_2 so that

$$F'(X) \subset f'(\xi) + p_1 d(X)\,[-1, 1]\,, \tag{5}$$

$$Q'(X) + I_\varepsilon \subset f'(\xi) + p_2 d(X)\,[-1, 1]\,. \tag{6}$$

From (3) and (6) we obtain

$$N_\varepsilon(X) \subset \alpha - \frac{f(\alpha)}{f'(\xi) + p_2 d(X)\,[-1, 1]}\,.$$

Hence

$$d(N_\varepsilon(X)) \leqq \frac{2p_2\,|f(\alpha)|\,d(X)}{M}\,, \tag{7}$$

where $M = f'(\xi)^2 - p_2^2 d(X)^2$.

If $d(X)$ is sufficiently small, then

$$|\alpha - \xi| = C_1 |a - \xi|^\lambda = C_2 |b - \xi|^\lambda ,$$

where C_1 and C_2 are finite positive constants. Since

$$|f(\alpha)| = |f(\xi) + (\alpha - \xi) f'(\xi + \theta(\alpha - \xi))| , \qquad \theta \in [0, 1] ,$$

on the basis of (5) we find

$$|f(\alpha)| \leqq \|F'(X)\| \cdot |\alpha - \xi| \leqq (|f'(\xi)| + p_1 d(X))\, d(X)^\lambda .$$

Using (7) we obtain

$$d(N_\varepsilon(X)) \leqq \frac{2p_2(|f'(\xi)| + p_1 d(X))\, d(X)^{\lambda+1}}{M} ,$$

i.e.

$$d(N_\varepsilon(X)) \leqq C d(X)^{\lambda+1} , \tag{8}$$

where C is a finite positive constant chosen so that

$$C \geqq \frac{2p_2(|f'(\xi)| + p_1 d(X))}{M} .$$

Choosing the initial interval $X_0 = X$ so that the inequality (8) is satisfied, we obtain

$$d(N_\varepsilon(X_0) \cap X_0) \leqq \min (d(N_\varepsilon(X_0)), d(X_0)) = (dN_\varepsilon(X_0)) \leqq$$
$$\leqq C d(X_0)^{\lambda+1} . \tag{9}$$

For the interval sequence (X_n) defined by (4) we have

$$X_{n+1} \subset X_n , \qquad d(X_{n+1}) \leqq d(N_\varepsilon(X_n)) .$$

According to mathematical induction, from (8) and (9) we obtain

$$d(X_{n+1}) \leqq C d(X)^{\lambda+1} , \qquad n = 0, 1, \dots ,$$

and Theorem is proved. □

Since λ is almost two (see [2]) the convergence order of the modified MOORE's method (4) is almost three.

We shall now consider *a procedure for the choice of the interval I_ε and the parameter ε*, assuming that $d(X)$ is sufficiently small. Let

$$E = \begin{cases} E_1 & \text{if } [E_1, E_2] > 0 \\ -E_2 & \text{if } [E_1, E_2] < 0 \end{cases} ,$$

and let ε_1 and ε_2 are taken from the set $\{0, \varepsilon\}$, where $\varepsilon < E$. Let us represent the interval I_ε in the form $I_\varepsilon = [-\varepsilon_1, \varepsilon_2]$.

Without loss of generality, we can assume that $f(\alpha) > 0$. From (3) we obtain

$$N_\varepsilon(X) = \left[\alpha - \frac{f(\alpha)}{E_1 - \varepsilon_1} , \quad \alpha - \frac{f(\alpha)}{E_2 + \varepsilon_2}\right] .$$

We have to choose the parameters ε_1 and ε_2 so that $\xi \in N_\varepsilon(X)$. This requirement is equivalent to the inequality

$$f\left(\alpha - \frac{f(\alpha)}{E_1 - \varepsilon_1}\right) f\left(\alpha - \frac{f(\alpha)}{E_2 + \varepsilon_2}\right) < 0 . \tag{10}$$

Since X is narrow, $f(\alpha)$ is a small quantity, so we can write

$$f(\alpha - t) \cong f(\alpha) - t f'(\alpha) + t^2 \frac{f''(\alpha)}{2} ,$$

where t takes the values $\dfrac{f(\alpha)}{E_1 - \varepsilon_1}$ and $\dfrac{f(\alpha)}{E_2 + \varepsilon_2}$. According to this, the inequality (10) becomes

$$\left[1 - \frac{f'(\alpha)}{E_1 - \varepsilon_1} + \delta_1\right]\left[1 - \frac{f'(\alpha)}{E_2 + \varepsilon_2} + \delta_2\right] < 0 , \tag{11}$$

where

$$\delta_1 = \frac{f''(\alpha) f(\alpha)}{2(E_1 - \varepsilon_1)^2} , \qquad \delta_2 = \frac{f''(\alpha) f(\alpha)}{2(E_2 + \varepsilon_2)^2} .$$

Since $\delta_1 = O(f(\alpha))$ and $\delta_2 = O(f(\alpha))$, δ_1 and δ_2 will not influence the sign of the expression on the left hand side of the inequality (11) if $\dfrac{f'(\alpha)}{E_1 - \varepsilon_1}$ and $\dfrac{f'(\alpha)}{E_2 + \varepsilon_2}$ are not close to 1, i.e. if

$$\left|\frac{E_1 - \varepsilon_1 - f'(\alpha)}{f(\alpha)}\right| \gg 1 , \quad \left|\frac{E_2 + \varepsilon_2 - f'(\alpha)}{f(\alpha)}\right| \gg 1 \tag{12}$$

holds.

Assume that $f'(\alpha) \in [E_1, E_2]$. If (12) is satisfied for $\varepsilon_1 = \varepsilon_2 = 0$, then the sign of (11) does not depend on δ_1 and δ_2. From (11) we obtain

$$\left(1 - \frac{f'(\alpha)}{E_1}\right)\left(1 - \frac{f'(\alpha)}{E_2}\right) < 0 ,$$

which is correct because $E_1 < f'(\alpha) < E_2$. Thus, if $f'(\alpha) \in [E_1, E_2]$ and (12) holds for $\varepsilon_1 = \varepsilon_2 = 0$, then $I_\varepsilon = [0, 0]$.

If (12) is not valid for $\varepsilon_1 = \varepsilon_2 = 0$, then we shall take $I_\varepsilon = [-\varepsilon, \varepsilon]$, where

$$\frac{E^*}{R} \leqq \varepsilon < E . \tag{13}$$

E^* is one of the numbers $|E_1|$, $|E_2|$, and R is a positive number such that $\dfrac{E^*}{R} = o(f(\alpha))$. It is desirable that (13) is satisfied for ε as small as possible.

Assume now that $f'(\alpha) \notin [E_1, E_2]$. If (13) holds, the quantities δ_1 and δ_2 will not influence the sign of the inequality (11). Therefore, it is sufficient to consider the simplified inequality

$$\left(1 - \frac{f'(\alpha)}{E_1 - \varepsilon_1}\right)\left(1 - \frac{f'(\alpha)}{E_2 + \varepsilon_2}\right) < 0 . \tag{14}$$

Since the function f is monotonical on X, we have $\operatorname{sgn}(f'(\alpha)) = \operatorname{sgn}([E_1, E_2])$. According to this, (13) and (14) we obtain the relations which determine the bounds of the parameters ε_1 and ε_2.

If $[E_1, E_2] > 0$, then:

(A) $f'(\alpha) < E_1 \Rightarrow E_1 > \varepsilon_1 > E_1 - f'(\alpha)$, $\varepsilon_2 > f'(\alpha) - E_2$;

(B) $f'(\alpha) > E_2 \Rightarrow E_1 > \varepsilon_1 > E_1 - f'(\alpha)$, $\varepsilon_2 > f'(\alpha) - E_2$.

If $[E_1, E_2] < 0$, then:

(C) $f'(\alpha) > E_2 \Rightarrow \varepsilon_1 > E_1 - f'(\alpha)$, $-E_2 > \varepsilon_2 > f'(\alpha) - E_2$;

(D) $f'(\alpha) < E_1 \Rightarrow \varepsilon_1 > E_1 - f'(\alpha)$, $-E_2 > \varepsilon_2 > f'(\alpha) - E_2$.

In the cases (B) and (C) we have $E_1 - f'(\alpha) < 0$ so that we can choose $\varepsilon_1 = 0$. Further, when $f'(\alpha) - E_2 < 0$ ((A) and (D)), we shall take $\varepsilon_2 = 0$. Regarding this, the relations (A), (B), (C), (D) and the mentioned remarks, the interval I_ε can be chosen as follows:

(A) $I_\varepsilon = [-\varepsilon, 0]$, $\quad \varepsilon \in \left(\max\left(E_1 - f'(\alpha), \dfrac{E_1}{R}\right), E_1\right)$;

(B) $I_\varepsilon = [0, \varepsilon]$, $\quad \varepsilon > \max\left(f'(\alpha) - E_2, \dfrac{E_2}{R}\right)$,

(C) $I_\varepsilon = [0, \varepsilon]$, $\quad \varepsilon \in \left(\max\left(f'(\alpha) - E_2, \dfrac{-E_2}{R}\right), -E_2\right)$;

(D) $I_\varepsilon = [-\varepsilon, 0]$, $\quad \varepsilon > \max\left(E_1 - f'(\alpha), \dfrac{-E_1}{R}\right)$.

Note that the lower bound for ε ensures the validity of (12).

The convergence order of the interval method (4) is increased but also the number of numerical operations. This decreases somewhat the efficience of the proposed method. On the other hand, an important advantage of the interval process (4) relative to MOORE's method and similar methods which use $F'(X)$ is its applicability under the weaker conditions. Namely, MOORE's process requires an initial interval being narrow enough to avoid division by a zero-interval. Applying the interval method (4), the requirement for the initial interval is not so strict. According to this, if the initial interval is not sufficiently narrow, the interval method (4) can be used for finding a contracted interval which is convenient to apply some other interval methods.

The mentioned remarks and the application of the interval process (4) are illustrated by the following example.

Example: The algebraic equation

$$f(x) \equiv x^5 - 9x^4 + 25x^3 - 15x^2 - 26x + 24 = 0$$

has a real simple zero ($\xi = 2$) in the interval $X_0 = [1.5, 2.3]$. In order to find a sufficiently narrow interval which contains the exact zero ξ of f, we shall first calculate the interval extension $F'(X_0)$. Using HORNER's rule, we find

$$F'(X_0) = (((5X_0 - 36)\,X_0 + 75)\,X_0 - 30)\,X_0 - 26 = \\ = [-62.3975,\ 107.3425]\,.$$

It is clear that $F'(X_0)$ is very wide. Besides, this interval contains the zero so that MOORE's method cannot be applied. Therefore, we shall apply the modified method [4]. The results of this application are shown in the table below.

	$n = 0$	$n = 1$	$n = 2$
a	1.5	1.988 764 195 3	1.999 999 999 9
b	2.3	2.002 875 314 1	2.000 000 001 5
A	−10.043 251 791 3	19.760 712 443 4	
B	3.585 360 005 2	−36.269 002 054 3	
k	0.509 591 404 6	−0.303 633 822 4	
α	2.021 487 056 1	2.000 000 156 1	
$f'(\alpha)$	5.947 371 072 3	5.999 999 688 3	
E_1	3.923 599 931 1	5.994 784 948 1	
E_2	5.898 409 312 6	6.020 525 386 2	
I_ε	[0, 1]	[0, 0]	

In the first iteration the interval $I_\varepsilon = [0, \varepsilon]$ $(\varepsilon = 1)$ is used (the case (B): $[E_1, E_2] > 0,\ f'(\alpha) > E_2$). In the second iteration we have chosen $I_\varepsilon = [0, 0]$ because $f'(\alpha) \in [E_1, E_2]$ and (12) holds for $\varepsilon_1 = \varepsilon_2 = 0$.

The width of the obtained intervals $X_0 = [1.5, 2.3]$, $X_1 = [1.9887641953, 2.0028753141]$ and $X_2 = [1.9999999999, 2.0000000015]$ are approximatively $d(X_0) = 8 \times 10^{-1}$, $d(X_1) \cong 1.4 \times 10^{-2}$, $d(X_2) \cong 1.6 \times 10^{-10}$. According to the values $d(X_i)$ we can conclude that the interval method (4) converges very fast.

References

1 MOORE, R. E., Interval Analysis, Prentice Hall, New Jersey 1966.
2 RIDDERS, C. J. F., Three-point iterations derived from exponential curve fitting, IEEE Trans. Circuits and Systems 8 (1979), 669–670.

Eingereicht: 16. 4. 1980

Anschrift: MIODRAG S. PETKOVIĆ, Faculty of Electronic Engineering, University of Niš, Beogradska 14, 18000 Niš, Yugoslavia

ZAMM 61, 119–120 (1981)

I. I. GLICK

A Note on Runs

Let S be a collection of $m \geqq 2$ objects. Suppose we sample with replacement from S infinitely many times. Let the first selection be made at time 0, the second at time 1, etc. We then get a sequence, $(x_0, x_1, x_2, \ldots)$, of outcomes. We assume that the probability of selecting any particular member of S at any given time is $p = 1/m$. Let W denote the space of all sequences of outcomes $w = (x_0, x_1, x_2, \ldots)$, i.e., $W = S \times S \times \ldots$, and let W be supplied with the product measure. In [1], L. RÅDE obtained the probability generating functions for the waiting times until the first occurrence of each of several patterns. One of those patterns is a run, $xx \ldots x$, which consists of k successive repetitions of the same element of S. Herein, by a different approach, namely considering a certain MARKOV chain, we will get RÅDE's result for this pattern, and at the same time, with no additional effort, we will get some probabilities and expected values connected with it that RÅDE did not mention.

Define the MARKOV *chain* $X_0, X_1, X_2, \ldots$ on W by $X_n(w) =$ the number of consecutive occurrences up through time n of whatever the outcome at time n is, e.g., if S consists of the three letters a, b, c, then X_4 $(abaaa \ldots) = 3$. The *state space* is the positive integers.

The *transition matrix* of the MARKOV chain is

$$\begin{bmatrix} 1-p & p & 0 & 0 & \cdots \\ 1-p & 0 & p & 0 & \cdots \\ 1-p & 0 & 0 & p & \cdots \\ \vdots & & \vdots & \vdots & \vdots \end{bmatrix}$$

Letting $p_{ij}^{(n)}$ denote the probability that starting from state i at some time the process is in state j n steps later, and using the usual conventions that $p_{ii}^{(0)} = 1$, and $p_{ij}^{(0)} = 0$ if $i \neq j$, we have

$$p_{1k}^{(n)} = \begin{cases} 0 & \text{if } 0 \leqq n \leqq k-2 \\ p^{k-1} & \text{if } n = k-1 \\ (1-p)\,p^{k-1} & \text{if } n \geqq k \end{cases},$$

and

$$p_{kk}^{(n)} = \begin{cases} 1 & \text{if } n = 0 \\ 0 & \text{if } 1 \leqq n \leqq k-1 \\ (1-p)\,p^{k-1} & \text{if } n \geqq k \end{cases}.$$

These can be calculated directly or read off from the powers of the transition matrix.

Let $k > 1$, and let $T(w)$ denote the least n such that $X_n(w) = k$. Let $T(w) = \infty$ if w does not contain k consecutive occurrences of any outome. T is then the first time that k consecutive occurrences of the same outcome happen. Letting N denote the number of performances of the experiment until the first time that k consecutive occurrences of the same outcome have happened, we have $N = 1 + T$. Let $f_{1k}^{(n)}$ be the probability that starting from state 1 the process enters state k for the first time afterward n steps later. $f_{1k}^{(n)} = \mathsf{P}(T = n \mid X_0 = 1)$. We note that $\mathsf{P}(X_0 = 1) = 1$, so

$$f_{1k}^{(n)} = \mathsf{P}(T = n \mid X_0 = 1) = \mathsf{P}(T = n)\,.$$

$f_{1k}^{(0)}$ is defined to be 0. Let $P_{1k}(s) = \sum\limits_{n=0}^{\infty} p_{1k}^{(n)} s^n$, $P_{kk}(s) = \sum\limits_{n=0}^{\infty} p_{kk}^{(n)} s^n$, and $F_{1k}(s) = \sum\limits_{n=0}^{\infty} f_{1k}^{(n)} s^n$. Then we have the well-known relation $P_{1k}(s) = F_{1k}(s)\,P_{kk}(s)$. Since

$$P_{1k}(s) = p^{k-1}s^{k-1} + (1-p)\,p^{k-1} \sum_{n=k}^{\infty} s^n \\ = \frac{p^{k-1}s^{k-1} - p^k s^k}{1-s},$$

and

$$P_{kk}(s) = 1 + (1-p)\,p^{k-1} \sum_{n=k}^{\infty} s^n \\ = \frac{1 - s + (1-p)\,p^{k-1}s^k}{1-s},$$

then

$$F_{1k}(s) = \frac{P_{1k}(s)}{P_{kk}(s)} = \frac{p^{k-1}s^{k-1} - p^k s^k}{1 - s + (1-p)\,p^{k-1}s^k}.$$

Note that $\sum\limits_{n=1}^{\infty} f_{1k}^{(n)} = F_{1k}(1) = 1$, so $P(T = \infty) = 0$.

$\sum\limits_{n=1}^{\infty} p_{kk}^{(n)} = P_{kk}(1) - 1 = \infty$, so state k is recurrent. States 1 and k communicate, and state k is recurrent, so

$$\mathsf{P}(\text{process starting in state 1 enters state } k \text{ infinitely often}) = 1\,.$$

Furthermore,

$$\mathsf{E}(T) = \sum_{n=0}^{\infty} n f_{1k}^{(n)} = F'_{1k}(1) = \frac{1 - p^{k-1}}{(1-p)\,p^{k-1}},$$

so

$$\mathsf{E}(N) = 1 + \mathsf{E}(T) = 1 + \frac{1 - p^{k-1}}{(1-p)\,p^{k-1}} = \frac{1 - p^k}{(1-p)\,p^{k-1}} \\ = \frac{m^k - 1}{m - 1},$$

which is RÅDE's result.

Finally, let $f_{kk}^{(n)}$ denote the probability that starting from state k the first return to state k occurs n steps later. Let $F_{kk}(s) = \sum\limits_{n=1}^{\infty} f_{kk}^{(n)} s^n$, the probability generating function. Then, again using a well-known result, we have

$$F_{kk}(s) = 1 - \frac{1}{P_{kk}(s)} = 1 - \frac{1-s}{1 - s + (1-p)\,p^{k-1}s^k} \\ = \frac{(1-p)\,p^{k-1}s^k}{1 - s + (1-p)\,p^{k-1}s^k}\,.$$

Then $F'_{kk}(1) = \frac{1}{(1-p)\,p^{k-1}} = \frac{m^k}{m-1}$. This is the expected waiting time for the first return to state k starting from state k. Then, if $r \geqq 1$, the expected waiting time until the rth entrance into state k is

$$\frac{m^k-1}{m-1} + \frac{(r-1)\,m^k}{m-1} = \frac{rm^k-1}{m-1}.$$

References

1 RÅDE, LENNART, Waiting for patterns in a sequence of random numbers, ZAMM 56 (1976), 124–125.

Eingereicht: 8. 4. 1980

Anschrift: Professor Dr. IRVING I. GLICK, The George Washington University, Department of Mathematics, Washington, D.C. 20052, USA.

ZAMM 61, 120–122 (1981)

R. M. PALAIYA / PRABHA MAJUMDAR

Interaction of Antiplane Shear Waves by Rigid Strip Lying at the Interface of two Bonded Dissimilar Elastic Half-Spaces

The problem of diffraction of normally incident SH waves by a rigid strip situated at the interface of two elastic half-spaces is solved. The problem is reduced to the solution of dual integral equations. Approximate solutions of the dual integral equation has been obtained by reducing it to a FREDHOLM integral equation of the second kind. The solutions have been used to calculate stress intensity factors at the edge of the strip.

1. Introduction

It is well-known that the problems of diffraction of elastic waves by cracks or inclusions are of considerable interest in seismology and geophysics. If the cracks or inclusions are located at the interface of layered media, the study becomes more relevant. The study of diffraction of elastic waves by cracks/inclusions in homogeneous, isotropic elastic solids has been the subject of many investigations. Recently SRIVASTAVA, PALAIYA and GUPTA [1], [2] have studied the interaction of longitudinal waves by a GRIFFITH penny-shaped crack located at the interface of two dissimilar elastic half-spaces. Corresponding problems for shear waves have been studied in [3] and [4].

In continuation to the above study, our object in this paper is to attempt the problem of diffraction of shear waves by a rigid strip situated at the interface of two dissimilar elastic half-spaces. Using FOURIER Cosine transforms, the problem has been reduced to the solution of a FREDHOLM integral equation of the second kind. An iterative solution, valid for smaller wave frequencies, has been obtained. These solutions are used to calculate stress intensity factors at the edge of the strip.

2. Formulation of the Problem

Let a rigid strip of infinite length and finite width be located at the interface of two bonded dissimilar, elastic semi-infinite solids. Consider a rectangular cartesian co-ordinate system (x_1, x_2, x_3) such that the strip occupies the region $-a \leqq x_1 \leqq a$, $-\infty < x_2 < \infty$, $x_3 = 0$ at the interface of two half-spaces $x_3 > 0$ and $x_3 < 0$. It is convenient to normalize all lengths with respect to a, half-width of the strip. Writing $x_1/a = x$, $x_2/a = y$, $x_3/a = z$, the strip is defined by relations $-1 \leqq x \leqq 1$, $-\infty < y < \infty$, $z = 0$ at the interface of the halfspaces $z \leqq 0$ and $z \geqq 0$. An antiplane shear wave is assumed to be incident normally on the strip. The only non-vanishing displacement is the y-direction component $u_y = u_y(x, z)\,\mathrm{e}^{-i\omega t}$. As a consequence, all stress components vanish identically except for the shear stress σ_{xy} and σ_{yz}. In the absence of body forces, the only equation of motion not identically satisfied by non-vanishing displacement is

$$(u_{y,xx} + u_{y,zz} + k^2 u_y)\,\mathrm{e}^{-i\omega t} = 0 \tag{2.1}$$

where $k = \omega a(\varrho/\mu)^{1/2}$, μ is the LAMÉ's constant, ϱ the material density, ω the circular wave frequency. It may be noted that k is the dimensionless wave frequency.

We have the problem of finding stress-distribution subject to the following boundary conditions:

$$u_y(x, 0+) = u_y(x, 0-) = -u_0(x)\,e^{-i\omega t}, \quad |x| \leqq 1, \tag{2.2}$$

$$\left.\begin{aligned} u_y(x, 0+) &= u_y(x, 0-) \quad (2.3) \\ \sigma_{yz}(x, 0+) &= \sigma_{yz}(x, 0-) \quad (2.4) \end{aligned}\right\} \text{ for } |x| > 1 .$$

In what follows, the factor $\mathrm{e}^{-i\omega t}$ shall be suppressed. Using FOURIER Cosine transforms, we get the solution of equation (2.1) suitable for the above problem:

$$u_y(x, z) = \begin{cases} \int_0^\infty A_1(\xi) \exp(-\beta_1 z) \cos \xi x \,\mathrm{d}\xi , & z \geqq 0 \\ \int_0^\infty A_2(\xi) \exp(\beta_2 z) \cos \xi x \,\mathrm{d}\xi , & z \leqq 0 \end{cases} \tag{2.5}$$

where

$$\beta_j = \begin{cases} (\xi^2 - k_j^2)^{1/2} , & k_j < \xi \\ -i(k_j^2 - \xi^2)^{1/2} , & k_j > \xi \end{cases} \tag{2.6}$$

$k_j^2 = \omega a(\varrho_j/\mu_j)^{1/2}$, $\quad j = 1, 2$.

The suffix 1 and 2 correspond to the half-spaces $z \geqq 0$ and $z \leqq 0$, respectively. In equation (2.5) $A_1(\xi)$ and $A_2(\xi)$ are unknown functions which have to be determined using the boundary conditions (2.2)–(2.4).

Corresponding to (2.5), we find the expressions for stress components:

$$\sigma_{yz}(x, z) = \begin{cases} -\mu_1 \int_0^\infty \beta_1 A_1(\xi) \exp(-\beta_1 z) \cos \xi x \,\mathrm{d}\xi , & z \geqq 0 \\ \mu_2 \int_0^\infty \beta_2 A_2(\xi) \exp(\beta_2 z) \cos \xi x \,\mathrm{d}\xi , & z \leqq 0 . \end{cases} \tag{2.7}$$

3. Derivation of the Integral Equation

From the boundary conditions (2.2) and (2.3) observe that $u_y(x, 0+) = u_y(x, 0-)$ for all values of x, hence we get

$$A_1(\xi) = A_2(\xi) . \tag{3.1}$$

The boundary conditions (2.2) and (2.4) lead to the following dual integral equations for the determination of the unknown function $A_1(\xi)$:

$$\int_0^\infty A_1(\xi) \cos \xi x \,\mathrm{d}\xi = -u_0(x) , \quad |x| < 1 , \tag{3.2}$$

$$\int_0^\infty (\mu_1\beta_1 + \mu_2\beta_2)\, A_1(\xi) \cos \xi x \,\mathrm{d}\xi = 0 , \quad |x| > 1 . \tag{3.3}$$

Writing

$$(\mu_1\beta_1 + \mu_2\beta_2)\,(\mu_1 + \mu_2)^{-1}\, A_1(\xi) = \Psi(\xi) \tag{3.4}$$

the above integral equations can be written in a convenient form

$$\int_0^\infty \xi^{-1}[1 + H(\xi)]\, \Psi(\xi) \cos \xi x \,\mathrm{d}\xi = -u_0(x) , \quad |x| < 1 , \tag{3.5}$$

$$\int_0^\infty \Psi(\xi) \cos \xi x \,\mathrm{d}\xi = 0 , \quad |x| > 1 \tag{3.6}$$

where

$$H(\xi) = \xi(\mu_1 + \mu_2)\,(\mu_1\beta_1 + \mu_2\beta_2)^{-1} - 1 . \tag{3.7}$$

It may be observed that $H(\xi) \to 0$ as $\xi \to \infty$.

In order to solve the dual integral equation (3.5–3.6) let us assume

$$\Psi(\xi) = \xi \int_0^1 y\theta(y)\, J_1(\xi y) \,\mathrm{d}y$$
$$= -\theta(1)\, J_0(\xi) + \int_0^1 \frac{\mathrm{d}}{\mathrm{d}y} \{y\theta(y)\}\, J_0(\xi y) \,\mathrm{d}y \tag{3.8}$$

where the function $\theta(y)$ shall soon be determined.

The equation (3.6) is automatically satisfied and (3.5) reduces to the FREDHOLM integral equation [5]:

$$\theta(y) + \int_0^1 \theta(t)\, G(t, y)\, dt = \frac{2}{\pi} \frac{d}{dy} \int_0^y \frac{u_0(x)}{(y^2 - x^2)^{1/2}}\, dx\,, \tag{3.9}$$

where

$$G(t, y) = t \int_0^\infty \xi H(\xi)\, J_1(\xi t)\, J_1(\xi y)\, d\xi\,. \tag{3.10}$$

The integrand in (3.10) has no poles, it has only branch points at $\xi = k_1$ and $\xi = k_2$. The infinite integral in equation (3.10) can be converted into integrals with finite limits by the procedure given in [6]. We shall give only the results:

Let $H(\xi) = L(\xi, \beta_1, \beta_2)$, *then*

$$G(t, y) = \frac{1}{2} t \left[\int_0^{k_1} \xi\{L(\xi, -i\beta_1', -i\beta_2') - L(\xi, i\beta_1', i\beta_2')\} \times H_1^{(1)}(\xi t) J_1(\xi y)\, d\xi + \int_{k_1}^{k_2} \xi\{L(\xi, \beta_1, -i\beta_2') - L(\xi, \beta_1, i\beta_2')\}\, H_1^{(1)}(\xi t)\, J_1(\xi y)\, d\xi \right], \quad t > y \tag{3.11}$$

where $\beta_j' = (k_j^2 - \xi^2)^{1/2}$, $j = 1, 2$ *and* $H_1^{(1)}(x)$, $J_1(x)$ *are the* BESSEL *functions.*

In the above derivations, we have taken $k_1 < k_2$. The value of this kernel for $k_1 > k_2$ is obtained by interchanging k_1 and k_2. For simplicity, however, we shall choose $k_1 < k_2$ in our further discussion. So also for $y > t$, t and y can be interchanged in (3.11).

On substituting the value for $L(\xi, \beta_1, \beta_2)$ in (3.11) from (3.7), we obtain

$$G(t, y) = it(\mu_1 + \mu_2)\, k_1^2 \left[\int_0^1 \frac{\xi^2 H_1^{(1)}(k_1 \xi t)\, J_1(k_1 \xi y)\, d\xi}{\mu_1(1 - \xi^2)^{1/2} + \mu_2(r^2 - \xi^2)^{1/2}} + \mu_2 \int_1^r \frac{\xi^2 (r^2 - \xi^2)^{1/2} H_1^{(1)}(k_1 \xi t)\, J_1(k_1 \xi y)\, d\xi}{\mu_1^2(\xi^2 - 1) + \mu_2^2(r^2 - \xi^2)} \right], \quad t > y \tag{3.12}$$

where $r = k_2/k_1 > 1$.

4. Solution of the Integral Equation

The dual integral equation (3.5)—(3.6) has been reduced to FREDHOLM integral equation (3.9) with the kernel $G(t, y)$ given by the equation (3.11). This form of the kernel is suitable for expansion in powers of k_1 (or k_2). Hence an iterative solution of (3.9), valid only for small values of the dimensionless frequency k_1 (*or* k_2) can be obtained.

For small values of the argument, the BESSEL functions $J_1(x)$ and $H_1^{(1)}(x)$ may be expanded in ascending powers of x as

$$H_1^{(1)}(x) = \left[1 + \frac{2i}{\pi} In\left(\frac{x}{2}\right)\right] J_1(x) + i \sum_{n=0}^{\infty} b_{2n+1} x^{2n-1}\,,$$

$$J_1(x) = \sum_{n=0}^{\infty} a_{2n+1} x^{2n+1}$$

where the coefficients a_{2n+1}, b_{2n+1} are given by ABRAMOVITZ and STEGUN [7]. Using the above expansions in (3.12), $G(t, y)$ may be expressed in the form

$$t^{-1} G(t, y) = k_1^2 G_1(t, y) + k_1^4 G_2(t, y) + k_1^4 \log k_1 G_3(t, y) + k_1^6 G_4(t, y) + k_1^6 \log k_1 G_5(t, y) + \ldots \tag{4.1}$$

where

$$G_1(t, y) = -a_1 b_1 y t^{-1} M_0\,,$$

$$G_2(t, y) = -\left(a_3 b_1 y^3 t^{-1} - i a_1^2 y t + a_1 b_3 y t + \frac{2 a_1^2}{\pi} y t \log \frac{t}{2}\right) M_2 - 2\pi^{-1} a_1^2 y t N_2\,,$$

$$G_3(t, y) = -2\pi^{-1} a_1^2 y t M_2\,,$$

$$G_4(t, y) = -\{-i a_1 a_3 y t (y^2 + t^2) + a_5 b_1 y^5 t^{-1} + a_3 b_3 y^3 t + a_1 b_5 y t^3 + 2\pi^{-1} a_1 a_3 y t (y^2 + t^2) \log (t/2)\}\, M_4 - 2\pi^{-1} a_1 a_3 y t (y^2 + t^2)\, N_4\,,$$

$$G_5(t, y) = -2\pi^{-1} a_1 a_3 y t (y^2 + t^2)\, M_4\,, \qquad y > t \tag{4.2}$$

and

$$M_n = \int_0^1 \alpha(\xi)\, d\xi + \int_1^r \beta(\xi)\, d\xi\,,$$

$$N_n = \int_0^1 \alpha(\xi) \log \xi\, d\xi + \int_1^r \beta(\xi) \log \xi\, d\xi\,, \tag{4.3}$$

$$\alpha(\xi) = (1 + \mu)\, \xi^{n+2} [\mu(1 - \xi^2)^{1/2} + (r^2 - \xi^2)^{1/2}]^{-1}\,,$$

$$\beta(\xi) = (1 + \mu)\, \xi^{n+2} (r^2 - \xi^2)^{1/2}\, [\mu^2(\xi^2 - 1) + (r^2 - \xi^2)]^{-1}\,,$$

$$\mu = \mu_1/\mu_2\,.$$

Next let us assume the solution of (3.9) in the form

$$\theta(y) = \theta_0(y) + k_1^2 \log k_1 \theta_1(y) + k_1^2 \theta_2(y) + (k_1^2 \log k_1)^2\, \theta_3(y) + k_1^4 \log k_1 \theta_4(y) + k_1^4 \theta_5(y) + o(k_1^6)\,. \tag{4.4}$$

Substituting this expansion as well as the expansion (4.1) of $G(t, y)$ in the integral equation (3.9) and equating the coefficients of equal powers of k_1, we get

$$\theta_0(y) = \frac{2}{\pi} \frac{d}{dy} \int_0^y \frac{u_0(x)}{(y^2 - x^2)^{1/2}}\, dx\,,$$

$$\theta_1(y) = \theta_3(y) = 0\,,$$

$$\theta_2(y) = -\int_0^1 t \theta_0(t)\, G_1(t, y)\, dt\,,$$

$$\theta_4(y) = -\int_0^1 t \theta_0(t)\, G_3(t, y)\, dt\,,$$

$$\theta_5(t) = -\int_0^1 t[\theta_2(t)\, G_1(t, y) + \theta_0(t)\, G_2(t, y)]\, dt\,.$$

In a particular, if we choose $u_0(x) = \frac{\pi}{2} c(1 - x)$,

c being a very small known constant, it can easily be seen that

$$\theta_0(y) = -c\,,$$

$$\theta_2(y) = -a_1 b_1 M_0 c y \left(1 - \frac{2}{3} y\right),$$

$$\theta_4(y) = -\frac{2 c a_1^2 M_2 y}{3\pi}\,, \quad \text{etc.}$$

For most practical cases, the dimensionless frequency k_1 is taken less than 0.6. For these values, (4.4) gives approximate solutions to a sufficient degree of accuracy.

5. Stress Intensity Factor

The stress component has a singularity at the edge of the strip. The singular part of the stress component can be obtained by using equations (2.7), (3.4) and (3.8).

Hence for $|x| < 1$, we have

$$\sigma_{yz}(x, 0+) = \sigma_{yz}(x, 0-) = -\mu_1 \int_0^1 \Psi(\xi) \cos \xi x\, d\xi + O(1)$$

$$= -\mu_1 \theta(1) \int_0^\infty J_0(\xi) \cos (\xi x)\, d\xi + O(1) =$$

$$= -\mu_1 \theta(1)\, (1 - x^2)^{-1/2} + O(1)\,, \quad |x| < 1\,. \tag{5.1}$$

Defining the stress intensity factor by the relation

$$N_z = \lim_{x \to 1-} (1 - x)^{1/2}\, |\sigma_{yz}(x, 0)|_{x<1}\,. \tag{5.2}$$

Using (5.1), we obtain

$$N_z = \frac{\mu_1}{\sqrt{2}}\, |\theta(1)|\,. \tag{5.3}$$

Solutions derived in section 4 may be used to calculate the values of the stress intensity factor for different combinations of the two elastic media and for a range of circular wave frequency. We have made calculations for the following values of the material constants which correspond to steel and aluminium respectively:

$\varrho_1 = 7.6$ gm/cm^3 , $\mu_1 = 8.32 \times 10^{11}$ dyne/cm^2 ,

$\varrho_2 = 2.7$ gm/cm^3 , $\mu_2 = 2.63 \times 10^{11}$ dyne/cm^2 .

The results are given in the following table:

k_1	0	0.1	0.2	0.3	0.4	0.5	0.6
$c^{-1}\|\theta(1)\|$	1.000	0.9996	0.9981	0.9941	0.9872	0.9815	0.9683

It may be observed that the stress-intensity factor decreases with the increase of the dimensionless frequency. This is in contrast with the results derived by SRIVASTAVA, PALAIYA and KAROLIA [3] in the case of GRIFFITH crack.

References

1 *Srivastava, K. N., Gupta, O. P.*, and *Palaiya, R. M.*, Interaction of elastic waves in two bonded dissimilar elastic half-planes having *Griffith* crack at interface, Int. J. Fracture, 14(2), 145 (1978).
2 *Srivastava, K. N., Palaiya, R. M.*, and *Gupta, O. P.*, Interaction of elastic waves with a penny-shaped crack at the interface of two bonded elastic half-spaces, Int. J. Fracture, 15(5), 591 (1979).
3 *Srivastava, K. N., Palaiya, R. M.*, and *Karolia, D. S.*, Interaction of antiplane shear waves by *Griffith* crack at the interface of two bonded dissimilar elastic half-planes, Int. J. Fracture (to appear).
4 *Srivastava, K. N., Palaiya, R. M.*, and *Karolia, D. S.*, Interaction of torsional waves by penny-shaped crack at the interface of two bonded dissimilar elastic half-spaces, ZAMM (to appear)
5 *Sneddon, I. N.*, Mixed boundary value problems in potential theory, North Holland Publishing Company, (1966).
6 *Mal, A. K.*, Interaction of elastic waves with a *Griffith* crack, Int. J. Engng. Sci. 8, 763 (1970).
7 *Abramowitz, M.*, and *Stegun, I. A.*, Handbook of Mathematical Functions, Dover N.Y. (1965).

Eingereicht am 31. 3. 1980

Anschrift: Dr. R. M. PALAIYA, Ass. Prof., Maulana Azad College of Technology, Bhopal-462007, India

ZAMM 61, 122 – 124 (1981)

K. MANIVACHAKAN / A. CHAKRABARTI

Application of a Generalised Fourier Transform to a Crack Problem

The formulation of a well-known crack problem is effected through the use of a Generalised FOURIER Transform introduced by the authors recently and the solution of the resulting singular integral equation is obtained easily by making use of POINCARÉ-BERTRAND formula. By this approach, dual integral equations formulation and the complicated procedure of their solution in terms of BESSEL functions are avoided. Two different pressure distributions along the crack surface leading to the same constant total pressure are considered as an application of the present approach to such crack problems.

1. Introduction

The problem of determining the distribution of stress in the vicinity of a GRIFFITH crack in an infinite plane when the crack is opened up by the application of a prescribed variable pressure to its surfaces has been solved through FOURIER transform method by SNEDDON and ELLIOTT [1] (see also [2] and [3]). In this method*), the boundary conditions are expressed as a set of dual integral equations involving the transform of the normal stress component and the same is solved by a complicated procedure in terms of BESSEL functions (see [2], p. 65). For this type of dual integral equations approach to other crack problems in classical Elasticity, the reader is also referred to [4]. In the present paper, use is made of a Generalised FOURIER transform (GFT) introduced by the authors recently [7] on the lines of GOLECKI [6] and the problem is reduced to that of solving a singular integral equation. The solution of this latter equation is readily available in literature and can also be obtained directly by making use of the POINCARÉ-BERTRAND formula. The solution so obtained directly gives the unknown stress component outside the crack surface. In this paper, we consider two different pressure distributions along the crack surface leading to the same constant total pressure and we show that the present method works well even when the assumed boundary stresses involve singularities of logarithmic type, in which case one finds it rather difficult to obtain a solution in closed form as obtained in this paper, using the approach of SNEDDON [3].

The method of this paper is well-suited for attacking elastostatic boundary value problems through SNEDDON's representation of the displacement vector in terms of harmonic functions and can be employed whether or not the displacement components are FOURIER-transformable, provided, of course, transforms exist for their first order derivatives and hence for the components of stress.

*) It is assumed that *Fourier* transforms exist for stress as well as displacement components.

2. Statement of the Problem

Assuming that the pressure applied is the same for both faces of the crack, which occupies the region $|x| < 1$ on $y = 0$, the problem considered is equivalent to that of finding the stress distribution in the half-plane $y \geqq 0$ when the boundary $y = 0$ is subjected to the conditions:

$$\sigma_{zy} = 0\,, \qquad -\infty < x < \infty \tag{2.1}$$

$$\sigma_{yy} = -r(x)\,, \qquad |x| < 1 \tag{2.2}$$

$$u_y = 0\,, \qquad |x| > 1 \tag{2.3}$$

where the function $r(x)$ is assumed to be even and prescribed as follows in two different situations:

Problem (1)

$$\left.\begin{aligned} r(x) &= \frac{P_0 - p}{\alpha} \quad \text{for } 0 < x < \alpha (0 < \alpha < 1) \\ &= P\delta(x - \eta) \quad \text{for } \alpha < x < 1 (\alpha < \eta < 1) \end{aligned}\right\} \tag{2.4}$$

(p_0 and P are constants. $\delta(x)$ is the DIRAC delta function).

Problem (2)

$$r(x) = p_0 - \frac{2p_0}{\pi}\,\frac{1}{(1 - x^2)^{1/2}} - \frac{p_0 x}{\pi(1 - x^2)^{1/2}} \ln\left(\frac{1 - x}{1 + x}\right). \tag{2.5}$$

Set

$$\sigma_{yy}(x, 0) = -q(x) \quad \text{for } |x| > 1 \tag{2.6}$$

where $q(x)$ is an even function of x. Thus we can write

$$\sigma_{yy}(x, 0) = -p(x)\,, \qquad -\infty < x < \infty \tag{2.7}$$

where $p(x)$ is an even function of x described by

$$\begin{aligned} p(x) &= r(x)\,, \quad 0 < x < 1\,, \\ &= q(x)\,, \quad x > 1\,. \end{aligned} \tag{2.8}$$

It is proposed to find $q(x)$ for the above two cases.

3. Formulation Using a GFT

We define a GFT $F(\xi, y)$ with respect to x of $f(x, y)$ by means of the following two equations:

$$-i\xi F = \int_{-\infty}^{\infty} \frac{\partial f}{\partial x}\, e^{i\xi x}\, dx\,, \tag{3.1}$$

$$\frac{\partial F}{\partial y} = \int_{-\infty}^{\infty} \frac{\partial f}{\partial y}\, e^{i\xi x} dx\,. \tag{3.2}$$

The inversion formulae are therefore given by

$$\frac{\partial f}{\partial x} = \frac{1}{2\pi} \int_{-\infty}^{\infty} (-i\xi F)\, e^{-i\xi x}\, d\xi\,, \tag{3.3}$$

$$\frac{\partial f}{\partial y} = \frac{1}{2\pi} \int_{-\infty}^{\infty} \frac{\partial F}{\partial y} e^{-i\xi x} d\xi . \tag{3.4}$$

Here we have assumed that $\partial f/\partial x$ and $\partial f/\partial y$ are absolutely integrable and absolutely continuous in $-\infty < x < \infty$.

According to SNEDDON [5], a solution to the equations of elastic equilibrium in the two-dimensional case may be obtained in terms of two functions φ and χ, harmonic in the plane, by writing the displacement components $u_x(x, y)$ and $u_y(x, y)$ as follows:

$$u_x = \frac{\partial \chi}{\partial x} + \frac{\partial \varphi}{\partial x} + (\beta^2 - 1)\, y \frac{\partial^2 \varphi}{\partial y\, \partial x}, \tag{3.5}$$

$$u_y = \frac{\partial \chi}{\partial y} - \beta^2 \frac{\partial \varphi}{\partial y} + (\beta^2 - 1)\, y \frac{\partial^2 \varphi}{\partial y^2} \tag{3.6}$$

where $\beta^2 = \frac{2(1-\sigma)}{1-2\sigma}$, σ being the POISSON ratio.

Let $U_x(\xi, y)$ and $U_y(\xi, y)$ be the GFT w.r.t. x of u_x and u_y, respectively, as defined by (3.1) and (3.2) and let Φ and X be the GFT w.r.t. x of φ and χ respectively as defined by GOLECKI [6] by means of the following equations:

$$-\xi^2 G = \int_{-\infty}^{\infty} \frac{\partial^2 g}{\partial x^2} e^{i\xi x} dx , \tag{3.7}$$

$$\frac{\partial^2 G}{\partial y^2} = \int_{-\infty}^{\infty} \frac{\partial^2 g}{\partial y^2} e^{i\xi x} dx , \tag{3.8}$$

$$-i\xi \frac{\partial G}{\partial y} = \int_{-\infty}^{\infty} \frac{\partial^2 g}{\partial x\, \partial y} e^{i\xi x} dx . \tag{3.9}$$

Transforming u_x and u_y with the help of (3.1), (3.5) and (3.6), one obtains:

$$U_x = (-i\xi)\left[X + \Phi + (\beta^2 - 1)\, y \frac{\partial \Phi}{\partial y}\right], \tag{3.10}$$

$$U_y = \frac{\partial X}{\partial y} - \beta^2 \frac{\partial \Phi}{\partial y} + (\beta^2 - 1)\, y \frac{\partial^2 \Phi}{\partial y^2} . \tag{3.11}$$

Again transforming φ with the help of (3.7) and (3.8) (Note that φ satisfies the equation $\partial^2\varphi/\partial x^2 + \partial^2\varphi/\partial y^2 = 0$) and solving the transformed differential equation, one obtains

$$\Phi(\xi, y) = A(\xi)\, e^{-|\xi| y} \tag{3.12}$$

taking into account the condition of vanishing of stresses as $y \to \infty$.

In a similar manner,

$$X = B(\xi)\, e^{-|\xi| y} . \tag{3.13}$$

To determine $A(\xi)$ and $B(\xi)$ we make use of the boundary conditions (2.1) and (2.7) in their transformed form

$$\bar\sigma_{xy}(\xi, 0) = 0; \qquad \bar\sigma_{yy}(\xi, 0) = -\bar p(\xi)$$

and obtain, using the stress-strain relations,

$$A(\xi) = \frac{(1+\sigma)\,(1-2\sigma)}{E\xi^2}\, \bar p(\xi) , \tag{3.14}$$

$$B(\xi) = 0 \tag{3.15}$$

where

$$\bar f(\xi) = \int_{-\infty}^{\infty} f(x)\, e^{i\xi x} dx$$

and E is the YOUNG's modulus.

On applying the inversion formula (3.3) to (3.11), we get, after substituting for Φ and X,

$$\frac{\partial u_y}{\partial x} = \frac{-2(1+\sigma)}{\pi E} \int_0^{\infty} P_c(\xi)\, [y\xi + 2(1-\sigma)]\, e^{-\xi y} \sin \xi x\, d\xi . \tag{3.16}$$

The boundary condition (2.3) which is equivalent to

$$\frac{\partial u_y}{\partial x} = 0 \quad \text{for } |x| > 1 \text{ on } y = 0$$

now yields

$$\int_0^{\infty} P_c(\xi)\, [y\xi + 2(1-\sigma)]\, e^{-\xi y} \sin \xi x\, d\xi = 0 \tag{3.17}$$

on $y = 0$ for $|x| > 1$.

Also

$$P_c(\xi) = \int_0^1 r(z) \cos \xi z\, dz + \int_1^{\infty} q(t) \cos \xi t\, dt . \tag{3.18}$$

We insert this expression in (3.17) and change the order of integration, evaluate the integrals for $y > 0$ (cf. [11]) and then take limits as $y \to 0$. We finally obtain

$$\int_1^{\infty} \frac{q(t)\, dt}{t^2 - x^2} = \int_0^1 \frac{r(z)\, dz}{x^2 - z^2} \quad \text{for } |x| > 1 . \tag{3.19}$$

This is a singular integral equation of the first kind for the unknown function $q(t)$, which represents the normal stress outside the crack on $y = 0$.

4. Solution of the Singular Integral Equation

Making the substitutions $t^2 = v^{-1}$ and $x^2 = u^{-1}$ in (3.19), we obtain

$$\int_0^1 \frac{L(v)\, dv}{v - u} = 2 \int_0^1 \frac{r(z)\, dz}{uz^2 - 1} = f(u) \quad \text{(say)} \quad (0 < u < 1) \tag{4.1}$$

where $q(t) = \sqrt{v}\, L(v)$.

Equation (4.1) is a singular integral equation of the first kind for $L(v)$ whose solution is available in [8] and [9]. However to solve the integral equation quickly, we may make use of the POINCARÉ-BERTRAND formula (cf. [10]) directly and obtain the solution as:

$$L(v) = \frac{C}{\pi \sqrt{v(1-v)}} - \frac{1}{\pi^2} \frac{1}{\sqrt{v\,(1-v)}} \int_0^1 \frac{\sqrt{u(1-u)}\, f(u)\, du}{u - v} \tag{4.2}$$

where

$$C = \int_0^1 L(v)\, dv.$$

The solution of (3.19) is therefore given by

$$q(t) = \frac{Ct}{\pi \sqrt{t^2 - 1}} - \frac{t}{\pi^2 \sqrt{t^2 - 1}} \int_0^1 \frac{\sqrt{u(1-u)}}{(u - t^{-2})} f(u)\, du \tag{4.3}$$

where

$$f(u) = 2 \int_0^1 \frac{r(z)\, dz}{uz^2 - 1}. \tag{4.4}$$

5. Solution for Stress and Displacement Field on $y = 0$

Let us now consider the two cases of pressure distribution $r(x)$ on the crack surface, as proposed in the Introduction.

Problem 1

For this case, using (2.4) and (4.4), we have

$$f(u) = \frac{-(p_0 - P)}{\alpha u^{1/2}} \ln\left(\frac{1 + \alpha u^{1/2}}{1 - \alpha u^{1/2}}\right) - \frac{2P}{1 - u\eta^2} .$$

Therefore,

$$q(t) = \frac{c}{\pi(1-v)^{1/2}} + \frac{1}{\pi^2(1-v)^{1/2}} \left[\frac{p_0 - P}{\alpha} I_1 + 2P I_2\right] \tag{5.1}$$

where

$$I_1 = \int_0^1 \frac{(1-u)^{1/2}}{u-v} \ln\left(\frac{1+\alpha u^{1/2}}{1-\alpha u^{1/2}}\right) du\,,$$

$$I_2 = \int_0^1 \frac{[u(1-u)]^{1/2}}{(u-v)(1-u\eta^2)} du\,.$$

The integrals I_1 and I_2 are evaluated using standard results (see [12], p. 81, prob. 18) and we obtain:

$$I_1 = \frac{\pi}{\alpha}\left[-2 + \frac{\pi\alpha(1-v)^{1/2}}{2} + \alpha(1-v)^{1/2} \arcsin \times \right.$$
$$\left. \times \left(\frac{(2-v)\,\alpha^2 - 1}{1 - v\alpha^2}\right) + 2(1-\alpha^2)^{1/2}\right], \tag{5.2}$$

$$I_2 = \frac{\pi}{\eta^2 v - 1}\left[v - \frac{1-(1-\eta^2)^{1/2}}{\eta^2}\right]. \tag{5.3}$$

The constant c in (5.1) can be determined using the condition $q(t) \to 0$ as $t \to \infty$. We, therefore, arrive at:

$$q(t) = \left(\frac{p_0 - P}{\alpha}\right)\left[\frac{1}{2} - \frac{2}{\pi}(\arcsin\alpha)\frac{t}{(t^2-1)^{1/2}} + \right.$$
$$\left. + \frac{1}{\pi}\arcsin\left(\frac{(2\alpha^2-1)\,t^2 - \alpha^2}{t^2-\alpha^2}\right)\right] + \frac{2Pt(1-\eta^2)^{1/2}}{\pi(\eta^2-t^2)\,(t^2-1)^{1/2}}\,,$$
$$(t > 1)\,. \tag{5.4}$$

It is observed that the approach of Sneddon [3] also gives the same result in this case. As a special case, consider a uniform pressure distribution

$$\sigma_{yy}(x, 0) = -p_0 \quad \text{for } |x| < 1 \tag{5.5}$$

on the surface of the crack.

From (5.4), we have, on putting $\alpha = 1$ and $P = 0$,

$$q(x) = -p_0\left[\frac{x}{(x^2-1)^{1/2}} - 1\right] \tag{5.6}$$

which agrees with that of Sneddon [3, p. 49].

In this case, we can also compute the normal displacement on $y = 0$ für $0 < x < 1$ using our GFT method. We have, from (3.16), (3.18), (5.5) and (5.6),

$$\frac{\partial u_y}{\partial x} = -2(1+\sigma)\,p_0(\pi E)^{-1} \cdot \int_0^\infty \left[[\xi^{-1}\sin\xi + \right.$$
$$\left. + \int_1^\infty 1 - x(x^2-1)^{-\frac{1}{2}} \cos\xi x\, dx\right][y\xi + 2(1-\sigma)]\, e^{-\xi y} \sin\xi x\, d\xi\,. \tag{5.7}$$

Computing the quantities on the right for $y > 0$ and then taking limits as $y \to 0$, we obtain

$$\frac{\partial u_y}{\partial x} = -2p_0(1-\sigma^2)\,E^{-1}x(1-x^2)^{-1/2} \quad \text{on } y = 0 \text{ for } |x| < 1\,.$$

Integrating w.r.t. x from x to 1 and noting that $u_y(1) = 0$,

$$u_y(x, 0) = 2p_0(1-\sigma^2)\,E^{-1}(1-x^2)^{1/2} \quad \text{for } 0 < x < 1\,. \tag{5.8}$$

This result also agrees with that of Sneddon [3, p. 49].

Problem 2

For this case,

$$f(u) = \frac{p_0}{u^{1/2}} \ln\left(\frac{1-u^{1/2}}{1+u^{1/2}}\right) + \frac{2p_0}{(1-u)^{1/2}} -$$
$$- \frac{p_0}{[u(1-u)]^{1/2}} \ln\left(\frac{1+u^{1/2}}{1-u^{1/2}}\right).$$

Therefore,

$$q(t) = \pi^{-1}Ct(t^2-1)^{-1/2} + \pi^{-2}p_0 t(t^2-1)^{-1/2}(I_3 - 2I_4 + I_5) \tag{5.9}$$

where

$$I_3 = I_1 \quad \text{with} \quad \alpha = 1\,,$$

$$I_4 = \int_0^1 u^{1/2}(u - t^{-2})^{-1}\, du\,,$$

$$I_5 = \int_0^1 (u - t^{-2})^{-1} \ln\left(\frac{1+u^{1/2}}{1-u^{1/2}}\right) du\,.$$

Using (5.2)

$$I_3 = -2\pi + \pi^2 t^{-1}(t^2-1)^{1/2}\,,$$

I_4 is easily evaluated (cf. [12, p. 9]) and is given by

$$I_4 = 2 + t^{-1}\ln\left(\frac{t-1}{t+1}\right),$$

$$I_5 = \int_0^1 \frac{1}{u-v}\left[\int_0^u \frac{ds}{[(u-s)(1-\delta s)]^{1/2}}\right] du =$$
$$= \int_0^1 \frac{1}{(1-s)^{1/2}}\left[\int_s^1 \frac{du}{(u-s)^{1/2}(u-v)}\right] ds$$

after changing the order of integration. The last integral is evaluated in a standard fashion and we obtain

$$I_5 = \frac{\pi^2}{2} - \frac{1}{2}\left[\ln\left(\frac{t-1}{t+1}\right)\right]^2.$$

Evaluating c as before, using the condition that $q(t) \to 0$ as $t \to \infty$, we finally arrive at:

$$q(t) = p_0\left\{1 - t(t^2-1)^{-1/2} - 2\pi^{-2}(t^2-1)^{-1/2}\ln\left(\frac{t-1}{t+1}\right) - \right.$$
$$\left. - 2^{-1}\pi^{-2}t(t^2-1)^{-1/2}\left[\ln\left(\frac{t-1}{t+1}\right)\right]^2\right\}(t > 1)\,. \tag{5.10}$$

Acknowledgement

The authors are grateful to the referee for his valuable suggestions.

References

1 *Sneddon, I. N.*, and *Elliott, H. A.*, The Opening of a *Griffith* Crack under Internal Pressure, Quart. App. Math., **4**, 262 (1946)
2 *Sneddon, I. N.*, *Fourier* Transforms. McGraw-Hill Book Company, Inc. (1951)
3 *Sneddon, I. N.*, The Use of Transform Method in Elasticity, Vol. I, North Carolina State University Reports AFOSR-64-1789 (1964)
4 *Sneddon, I. N.*, and *Lowengrub, M.*, Crack Problems in the Classical Theory of Elasticity, John Wiley & Sons, New York (1969)
5 *Sneddon, I. N.*, Solutions of the Equations of Thermoelastic Equilibrium, Archiwum Mech. Stos., **14**, 113–125 (1962)
6 *Golecki, J. J.*, Divergent Integral Appearing in *Fourier* Transform Applications in Elasticity, Int. J. Math. Educ. Sci. Technol., **6**, 321–326 (1975)
7 *Manivachakan, K.*, and *Chakrabarti, A.*, A Generalised *Fourier* Transform and its Use in *Flamant's* Problem of the Clasical Theory of Elasticity (to appear in Int. J. Math. Educ. Sci. Technol.)
8 *Peters, A. S.*, A Note on the Integral Equation of the First Kind with a *Cauchy* Kernel, Comm. Pure Appl. Maths., **16**, 57–61 (1963)
9 *Muskhelishvili, N. I.*, Singular Integral Equations, P. Noordhoff, Groningen, (1953)
10 *Baird, Sancaktar* and *Zweifel*, *Poincare-Bertrand* Formula, Siam J. Math. Anal., **8**, 580 (1977)
11 *Gradshteyn* and *Ryzhik*, Tables of Integrals, Series and Products, Academic Press, London (1965)
12 *Gakhov, F. D.*, Boundary Value Problems, Addison-Wesley Publishing Company, Inc., London (1966)

Eingereicht am 15. 7. 1979, in revidierter Fassung am 5. 5. 1980

Anschrift: Dr. K. Manivachakan, Department of Applied Mathematics, Indian Institute of Science, Bangalore – 560012, India

ZAMM 61, 125 – 126 (1981)

C. Y. WANG

The Ridging of Heavy Elastica

1. Introduction and Formulation

Suppose a thin, horizontal, heavy lamina is subjected to equal and opposite compressive horizontal forces. The lamina tends to buckle and a ridge is subsequently formed. This phenomenon occurs in the creasing of heavy fabrics such as carpeting. It is also important in the study of large geological ridges of the earth's crust.

We assume the lamina is thin compared to its other dimensions. In addition, we limit our study to deformations where the material can be considered linearly elastic. For large deflections the shape is governed by the heavy elasticy equations [1, 2]:

$$EI\,\frac{d^2\theta}{ds'^2} = -F' \sin\theta + \varrho(L' - s')\cos\theta\,, \tag{1}$$

$$\frac{dx'}{ds'} = \cos\theta\,, \qquad \frac{dy'}{ds'} = \sin\theta\,. \tag{2}$$

Here EI is the flexural rigidity, ϱ is the weight per length of the lamina, F' is the compressive force, $2L'$ is the total arc length of the raised ridge, x', y' are cartesian coordinates, s' is the arc length and θ is the local angle of inclination (Fig. 1). The boundary conditions are

$$\text{at}\quad s' = 0;\quad x' = y' = \theta = \frac{d\theta}{ds'} = 0\,, \tag{3}$$

$$\text{at}\quad s' = L'\,,\quad \theta = 0\,. \tag{4}$$

The unknowns are θ, x', y', and L'. We can normalize the variables as follows:

$$s = \sqrt{\frac{F'}{EI}}\,s'\,,\quad x = \sqrt{\frac{F'}{EI}}\,x'\,,\quad y = \sqrt{\frac{F'}{EI}}\,y'\,,\quad L = \sqrt{\frac{F'}{EI}}\,L'\,. \tag{5}$$

Eqs. (1, 2) become

$$\frac{d^2\theta}{ds^2} = -\sin\theta + \Lambda(L - s)\cos\theta\,, \tag{6}$$

$$\frac{dx}{ds} = \cos\theta\,, \qquad \frac{dy}{ds} = \sin\theta \tag{7}$$

where $\Lambda \equiv \varrho(EI)^{1/2}(F')^{-3/2}$ is a non-dimensional parameter. Eqs. (6, 7) are impossible to be solved in closed form.

2. The Approximate Solution for small θ

This is the case where the lamina is slightly buckled. Since $\theta \ll 1$ everywhere, Eq. (6) shows Λ must be small too. Let $\Lambda \equiv \varepsilon$ and we set

$$\theta = \varepsilon\varphi(s) + O(\varepsilon^3)\,, \tag{8}$$

$$y = \varepsilon\eta(s) + O(\varepsilon^3)\,, \qquad x = s + \varepsilon^2\xi(s) + O(\varepsilon^4)\,. \tag{9}$$

Eqs. (6, 7) give, as first approximations

$$\frac{d^2\varphi}{ds^2} = -\varphi + L - s\,, \tag{10}$$

$$\frac{d\xi}{ds} = -\frac{1}{2}\varphi^2\,, \qquad \frac{d\eta}{ds} = \varphi\,. \tag{11}$$

The boundary conditions are

$$\text{at}\quad s = 0;\quad \xi = \eta = \varphi = \frac{d\varphi}{ds} = 0\,, \tag{12}$$

$$\text{at}\quad s = L;\quad \varphi = 0\,. \tag{13}$$

Solving Eqs. (10, 12) we find

$$\varphi = \sin s - L\cos s + L - s\,. \tag{14}$$

Then Eq. (13) gives

$$\tan L = L\,,\quad \text{or } L = 4.49341\,. \tag{15}$$

The other roots of Eq. (15) give negative η and are discarded. Eqs. (11, 12) give

$$\eta = 1 - \cos s - L\sin s + Ls - \frac{1}{2}s^2\,, \tag{16}$$

$$\xi = -\frac{1}{2}\left[\frac{(L^2 - 1)}{4}\sin 2s + \frac{L}{2}\cos 2s + 2Ls\sin s + 2s\cos s - {} - 2(L^2 + 1)\sin s + \frac{1}{3}s^3 - Ls^2 + \left(\frac{1}{2} + \frac{3}{2}L^2\right)s - \frac{L}{2}\right]. \tag{17}$$

Let us define T as the normalized applied force:

$$T \equiv \frac{F'}{\varrho^{2/3}(EI)^{1/3}} = \Lambda^{-2/3}\,. \tag{18}$$

Define δ as the normalized horizontal displacement of the lamina due to T:

$$\delta \equiv \frac{2(L' - x'(L'))}{\varrho^{-1/3}(EI)^{1/3}} = 2T^{-1/2}(L - x(L))\,. \tag{19}$$

Define M as the maximum bending moment which occurs at $s = L$:

$$M \equiv \frac{EI\,\frac{d\theta}{ds'}(L')}{\varrho^{1/3}(EI)^{2/3}} = T^{1/2}\,\frac{d\theta}{ds}(L)\,. \tag{20}$$

Using Eqs. (4—20) we find, for small Λ or large T,

$$\delta = 75.6043\,T^{-7/2} + O(T^{-13/2})\,, \tag{21}$$

$$M = -5.60334\,T^{-1} + O(T^{-4})\,, \tag{22}$$

$$h \equiv \frac{y'(L')}{\varrho^{-1/3}(EI)^{1/3}} = 15.6987\,T^{-2} + O(T^{-5})\,, \tag{23}$$

$$w \equiv \frac{2x'(L')}{\varrho^{-1/3}(EI)^{1/3}} = 4.4934\,T^{-1/2} - 37.8022\,T^{-7/2} + O(T^{-13/2}) \tag{24}$$

where h is the normalized maximum height of the ridge and w is the normalized width of the ridge.

3. Exact Numerical Solution and Discussion

Numerical integration can be done by a one-parameter shooting method. Given $\Lambda = T^{-3/2}$ we guess L and integrate Eqs. (6, 7) by the fourth order RUNGE-KUTTA algorithm with zero initial conditions. A solution is obtained if $\theta = 0$ at $s = L$. If not we can adjust L, say, by NEWTON's method and integrate again. The results are shown in Fig. 2. We see that our approximate results, Eqs. (21—24), compare well with the exact numerical solution for $T > 4$.

Notice the width of the ridge w (the distance between the points of departure of the lamina with the horizontal) does not correspond to the distance shortened δ. The maximum width of the ridge is $w \approx 4.1$ occuring at $T \approx 3.7$.

Given a displacement δ one can obtain the force T required to maintain this displacement. Contrary to intuition, less force is required to maintain a larger displacement δ, since $\delta(T)$ has negative slope. Thus given a constant force T the initially horizontal lamina may not buckle at all ($\delta \to 0$, $T \to \infty$) but when the lamina is perturbed the ridge suddenly completely coalesces. This instability should be investigated experimentally.

Fig. 3 shows the shapes of ridges corresponding to different T. Coalescence occurs when $T \leqq 1.693$ or $\delta \geqq 5.117$. Actually gravitational instability may topple the configuration before δ reaches 5.117. Geologically failure occurs for much smaller δ, since the limiting criteria is probably the maximum moment M.

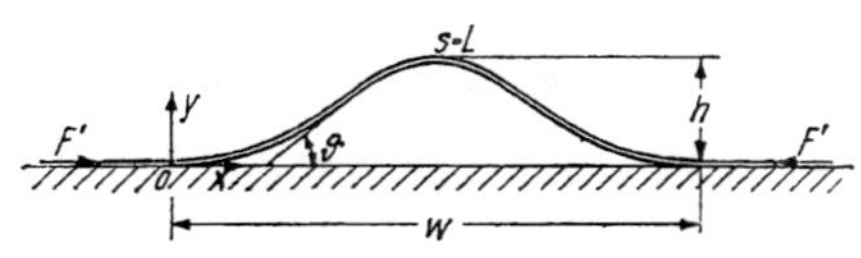

Fig. 1. The coordinate system

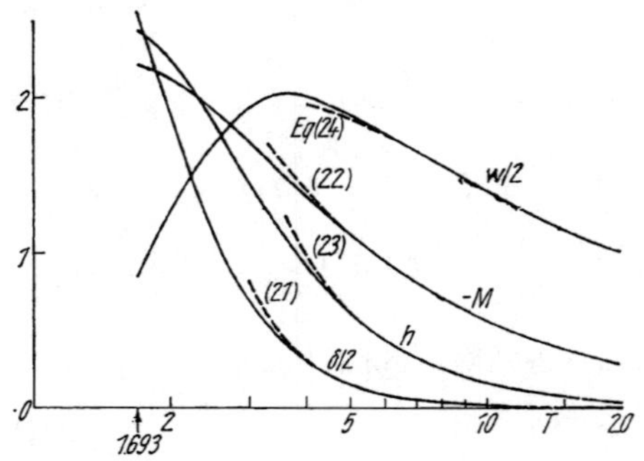

Fig. 2. Shortening δ, maximum moment M, width w, and height h as a function of force T. ——— exact numerical integration; – – – approximate solutions

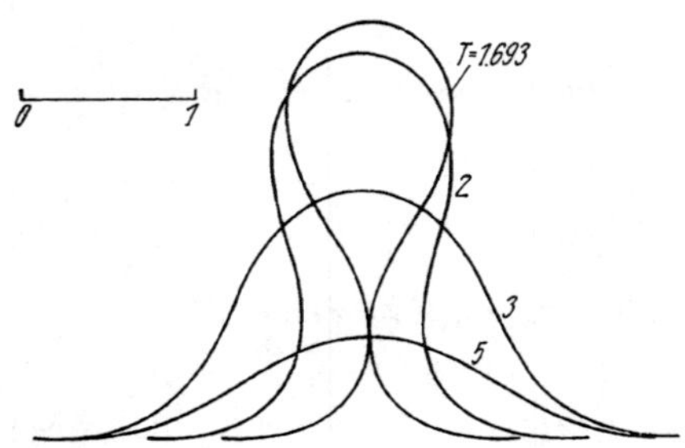

Fig. 3. Configurations of the ridge for various T

References

1 Bickley, W. G., The heavy elastica, Phil. Mag. Ser. 7, Vol. 17, 1934, pp. 603 to 622.

2 Frisch-Fay, R., Flexible Bars, Butterworths, Washington D.C., 1962, Chap. 2.

Eingereicht am 10. 4. 1980

Anschrift: Dr. C. Y. Wang, Professor, Michigan State University, Department of Mathematics, East Lansing, Michigan, 48824, USA

ZAMM 61, 126 – 127 (1981)

H. Bufler

Bemerkungen zum Bettischen Satz

Ein Reziprozitätssatz bezüglich der Größen $\boldsymbol{q} \in U$ und $\boldsymbol{u} \in U$ existiert immer dann, wenn ein Randwertproblem durch die Differentialgleichungen (bei einem diskreten System sind es algebraische Gleichungen)

$$\boldsymbol{A}\sigma = \boldsymbol{q} \tag{1}$$

$$\boldsymbol{B}\boldsymbol{u} = \boldsymbol{\varepsilon} \tag{2}$$

mit den linearen Operatoren $\boldsymbol{A} : \Sigma \to U$, $\boldsymbol{B} : U \to \Sigma$ sowie durch die lineare algebraische Beziehung

$$\boldsymbol{\varepsilon} = \boldsymbol{D}\sigma \tag{3}$$

mit $\boldsymbol{D} : \Sigma \to \Sigma$ beschrieben wird und sowohl $\boldsymbol{A}$ adjungiert zu $\boldsymbol{B}$ gemäß

$$[\boldsymbol{B}\boldsymbol{u}, \sigma] = \langle \boldsymbol{A}\sigma, \boldsymbol{u} \rangle \tag{4}$$

als auch $\boldsymbol{D}$ symmetrisch gemäß

$$[\boldsymbol{D}\sigma_1, \sigma_2] = [\boldsymbol{D}\sigma_2, \sigma_1] \tag{5}$$

ist (hierbei bedeutet $[\sigma, \varepsilon] = [\varepsilon, \sigma]$ das Skalarprodukt in Σ und $\langle \boldsymbol{q}, \boldsymbol{u} \rangle = \langle \boldsymbol{u}, \boldsymbol{q} \rangle$ das Skalarprodukt in U), siehe dazu [1]. Die Größe $\boldsymbol{q}$ kann dabei (auch nichtlinear) von $\boldsymbol{u}$ abhängen:

$$\boldsymbol{q} = \boldsymbol{C}(\boldsymbol{u}) . \tag{6}$$

Versieht man die Größen, welche (1) bzw. (2) genügen, mit dem Index 1 bzw. 2, so folgt nämlich aus (1) bis (5) sofort der Reziprozitätssatz

$$\langle \boldsymbol{q}_1, \boldsymbol{u}_2 \rangle = \langle \boldsymbol{q}_2, \boldsymbol{u}_1 \rangle . \tag{7}$$

Im Rahmen der linearen Elastizitätstheorie repräsentiert σ den Spannungstensor, ε den Verzerrungstensor, $\boldsymbol{u}$ den Verschiebungsvektor, $\boldsymbol{q}$ den eingeprägten Belastungsvektor und (3) das Hookesche Gesetz. (7) ist dort als Bettischer Satz bekannt. Gleichungen analoger Struktur treten auch bei anderen Problemen auf, z. B. bei der stationären Wärmeleitung: Hier ist σ der (negative) Wärmestromvektor, ε der Temperaturgradient, $\boldsymbol{u}$ die Temperatur, $\boldsymbol{q}$ die Wärmequellenbelegung und (3) die Wärmeleitungsgleichung.

Es ist nun interessant, festzustellen, daß auch in der geometrisch nichtlinearen Balkenstatik ein Reziprozitätssatz existiert. Wir betrachten hierzu (als Beispiel) den Balken gemäß Abb. 1 mit einem linearen Momenten-Krümmungs-Gesetz. Für ihn gelten folgende Gleichungen [2]:

$$\begin{bmatrix} -\dfrac{dM}{dx} \\ 0 \\ M(l) \end{bmatrix} = \begin{bmatrix} Q \\ 0 \\ 0 \end{bmatrix} \quad \begin{matrix} 0 < x < l \\ x = 0 \\ x = l \end{matrix} \tag{1'}$$

kurz $\boldsymbol{A}\boldsymbol{m} = \boldsymbol{q}$,

$$\begin{bmatrix} \dfrac{d\beta}{dx} \\ \beta(0) \\ 0 \end{bmatrix} = \begin{bmatrix} \varkappa \\ 0 \\ 0 \end{bmatrix} \quad \begin{matrix} 0 < x < l \\ x = 0 \\ x = l \end{matrix} \tag{2'}$$

kurz $\boldsymbol{B}\boldsymbol{\beta} = \boldsymbol{\varkappa}$,

$$\begin{bmatrix} \varkappa \\ 0 \\ 0 \end{bmatrix} = \begin{bmatrix} DM \\ 0 \\ 0 \end{bmatrix} \quad \begin{matrix} 0 < x < l \\ x = 0 \\ x = l \end{matrix} \tag{3'}$$

kurz $\boldsymbol{\varkappa} = \boldsymbol{D}\boldsymbol{m}$.

Ferner ist

$$[\boldsymbol{B}\boldsymbol{\beta}, \boldsymbol{m}] = \langle \boldsymbol{A}\boldsymbol{m}, \boldsymbol{\beta} \rangle , \tag{4'}$$

wobei

$$[\boldsymbol{B}\boldsymbol{\beta}, \boldsymbol{m}] = \int_0^l \frac{d\beta}{dx} M \, dx + \beta(0) \, M(0) ,$$

$$\langle \boldsymbol{A}\boldsymbol{m}, \boldsymbol{\beta} \rangle = -\int_0^l \frac{dM}{dx} \beta \, dx + M(l) \, \beta(l) .$$

Außerdem ist

$$[\boldsymbol{D}\boldsymbol{m}_1, \boldsymbol{m}_2] = [\boldsymbol{D}\boldsymbol{m}_2, \boldsymbol{m}_1] \quad \text{mit} \quad [\boldsymbol{D}\boldsymbol{m}_1, \boldsymbol{m}_2] = \int_0^l DM_1M_2 \, dx \tag{5'}$$

trivialerweise erfüllt. In obigen Formeln bedeuten

$$\boldsymbol{m} = \begin{bmatrix} M \\ M(0) \\ M(l) \end{bmatrix}, \quad \boldsymbol{\varkappa} = \begin{bmatrix} \varkappa \\ 0 \\ 0 \end{bmatrix}, \quad \boldsymbol{\beta} = \begin{bmatrix} \beta \\ \beta(0) \\ \beta(l) \end{bmatrix}, \quad \boldsymbol{q} = \begin{bmatrix} Q \\ 0 \\ 0 \end{bmatrix},$$

mit M als Biegemoment, $\varkappa$ als Krümmung, β als Querschnittsdrehung und Q als Querkraft. Die Querkraft ist mit β folgendermaßen verknüpft (Abb. 1):

$$\begin{bmatrix} Q \\ 0 \\ 0 \end{bmatrix} = \begin{bmatrix} C(\beta) \\ 0 \\ 0 \end{bmatrix} = \begin{bmatrix} (R + \int_x^l p(\xi) \, d\xi) \cos \beta \\ 0 \\ 0 \end{bmatrix}, \tag{6'}$$

kurz $\boldsymbol{q} = \boldsymbol{C}(\boldsymbol{\beta})$,

und die Durchbiegung errechnet sich aus

$$w(x) = -\int_0^x \sin \beta \, d\xi . \tag{8}$$

Gemäß (7) gilt nun der Reziprozitätssatz

$$\langle \boldsymbol{q}_1, \boldsymbol{\beta}_2 \rangle = \langle \boldsymbol{q}_2, \boldsymbol{\beta}_1 \rangle \quad \text{bzw.} \quad \int_0^l Q_1\beta_2 \, dx = \int_0^l Q_2\beta_1 \, dx . \tag{7'}$$

Gegenüber dem BETTIschen Satz in seiner üblichen Form kommt in (7′) die Querkraft Q anstelle der eingeprägten Kraft p und die Querschnittsdrehung β anstelle der Durchbiegung w vor. Um den Zusammenhang mit diesen Größen aufzuzeigen, nehmen wir daher eine geeignete Umformung vor.

Mit der Definition

$$f_i(x) = \int_x^l p_i(\xi)\,d\xi + R_i \qquad (i = 1, 2) \tag{9}$$

erhält man zunächst durch partielles Integrieren bei Beachtung von (8)

$$\int_0^l p_1(x)\,w_2(x)\,dx + R_1 w_2(l) = -\int_0^l f_1(x)\sin\beta_2(x)\,dx$$

und damit

$$\int_0^l f_1\cos\beta_1\beta_2\,dx = \int_0^l f_1(\cos\beta_1\beta_2 - \sin\beta_2)\,dx - \int_0^l p_1 w_2\,dx - R_1 w_2(l)\,. \tag{10}$$

Der Reziprozitätssatz (7′) geht mit (6′) und (10) nunmehr über in

$$\int_0^l p_1 w_2\,dx + R_1 w_2(l) = \int_0^l p_2 w_1\,dx + R_2 w_1(l) - r\,, \tag{11}$$

wobei

$$-r = \int_0^l f_1(\cos\beta_1\beta_2 - \sin\beta_2)\,dx - \int_0^l f_2(\cos\beta_2\beta_1 - \sin\beta_1)\,dx$$

$$= \int_0^l \left\{ f_1\beta_2\left(\frac{\beta_2^2}{6} - \frac{\beta_1^2}{2}\right) - f_2\beta_1\left(\frac{\beta_1^2}{6} - \frac{\beta_2^2}{2}\right) + \ldots \right\} dx\,. \tag{12}$$

Im Falle kleiner Neigungen der Balkenachse ($\beta_i \ll 1$) ist $r \approx 0$ und (11) entspricht dann dem klassischen BETTIschen Satz der linearen Theorie. Die dimensionslose Größe

$$\varrho = r \Bigg/ \left\{ \int_0^l p_2 w_1\,dx + R_2 w_1(l) \right\}$$

kann als Maß für die geometrische Nichtlinearität interpretiert werden.

Als numerisches Beispiel sei der Balken nach Abb. 2 betrachtet. Hierfür geht (11) über in

$$P_1 w_2(l_1) = P_2 w_1(l_2) - r$$

und es folgt für die Wahl $P_1 = P_2 = P$

$$\varrho = \frac{P_2 w_1(l_2) - P_1 w_2(l_1)}{P_2 w_1(l_2)} = \frac{w_1(l_2) - w_2(l_1)}{w_1(l_2)}\,.$$

In Abb. 3 sind die Durchbiegungen $w_1(l_2)$ (infolge der Belastung $P_1 = P$) und $w_2(l_1)$ (infolge der Belastung $P_2 = P$) in dimensionsloser Form aufgetragen. Aus Abb. 4 kann die Größe ϱ als Funktion von P abgelesen werden.

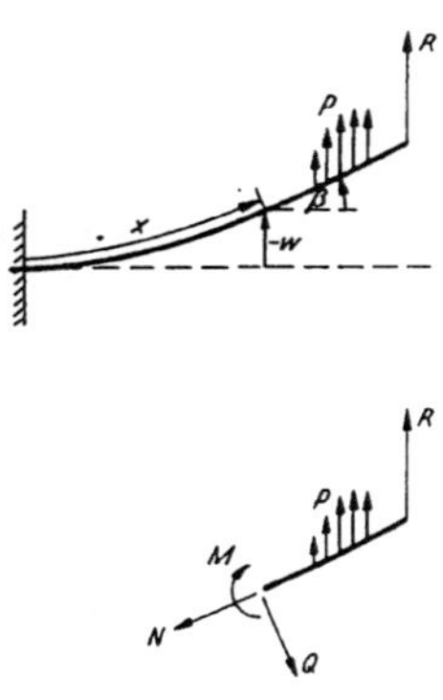

Abb. 1. Bezeichnungen am Balken

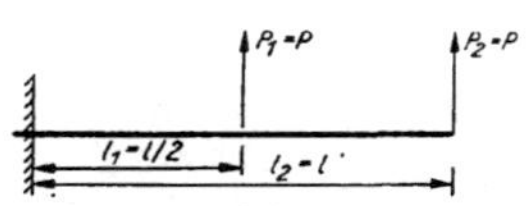

Abb. 2. Zum numerischen Beispiel

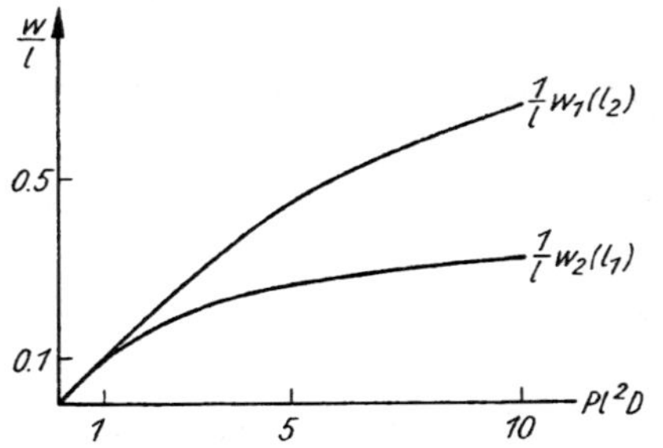

Abb. 3. Zum Balken der Abb. 2 gehörende Durchbiegungen

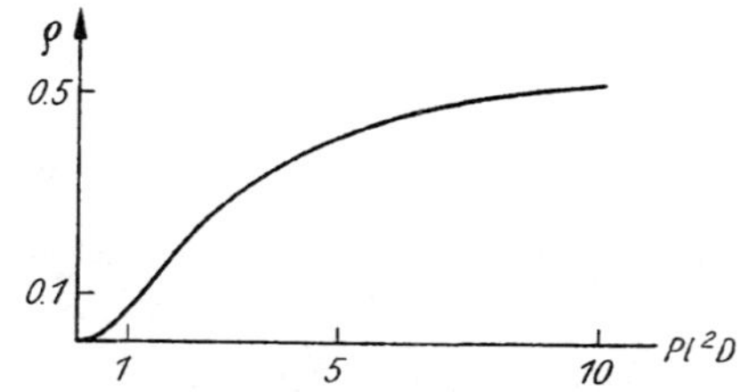

Abb. 4. Dimensionslose Größe ϱ in Abhängigkeit von der Belastung

Literatur

1 BUFLER, H., BETTIscher Satz, konservative Lasten und indifferentes Gleichgewicht, ZAMM 57, T 112 (1977).
2 BUFLER, H., Zur nichtlinearen Balkenstatik, ZAMM 60, T 108–T 112 (1980).

Eingereicht am 30. 4. 1980

Anschrift: Prof. Dr.-Ing. H. BUFLER, Institut für Mechanik (Bauwesen) der Universität Stuttgart, Pfaffenwaldring 7, 7000 Stuttgart 80, BRD

ZAMM 61, 127–129 (1981)

J. BAUMGARTE

Elimination holonomer Bindungen über generalisierte Koordinaten im Hamilton-Formalismus

Es soll die Frage aufgeworfen werden, ob man die Mechanik allein in der HAMILTON-Sprache behandeln kann, ohne auf die LAGRANGEsche Formulierung einzugehen. Es macht keine Schwierigkeiten, die HAMILTONfunktion bzw. die kinetische Energie für einzelne Massenpunkte aufzustellen. Das gleiche gilt für die kinetische Energie des starren Körpers. Denn aus der Kenntnis des Drehgeschwindigkeitsvektors im mitbewegten Bezugssystem als Funktion der EULERschen Winkel und ihrer Ableitungen kann man sofort auf die HAMILTONsche Formulierung übergehen. Ist nun aber das System holonomen Bindungen unterworfen, und will man diese Bindungen über generalisierte Variable in der HAMILTONfunktion erfüllen (Reduktion der Freiheitsgrade für jede holonome Nebenbedingung um eins, d. h. Reduktion der Ordnung um zwei), so ergeben sich im Gegensatz zur LAGRANGEschen Formulierung Schwierigkeiten. H. VOLZ diskutiert in seinem Buch [1] diese Schwierigkeiten und kommt dabei zum Schluß, daß seiner Meinung nach eine „Physikalische Nichtäquivalenz zwischen dem LAGRANGEschen und dem HAMILTONschen Formalismus" besteht.

Zur Klärung dieser Problematik zeigen wir zunächst, wie man in der LAGRANGEschen Formulierung die Reduktion auf generalisierte Koordinaten vornimmt, um nachher die doch existierende Äquivalenz zwischen dem HAMILTONschen und LAGRANGEschen Formalismus zu zeigen. Mit der LAGRANGEfunktion L, den Nichtpotentialkräften Q_i, den r holonomen Nebenbedingungen $f_\varrho = 0$ und den LAGRANGEschen Multiplikatoren λ_ϱ folgen die Bewegungsgleichungen (beschrieben in einem Inertialsystem) aus dem Variationsproblem:

$$\int_{t_0}^{t_1} dt\{\delta[L(q_i, \dot q_i, t) + \sum_{\varrho=1}^{r} \lambda_\varrho f_\varrho(q_i, t)] + \sum_i Q_i(q_j, \dot q_j, t)\, \delta q_i\} = 0\,,$$

$$L = \tfrac{1}{2} \sum_{i,k} a_{ik}\dot q_i \dot q_k - V(q_i, t)\,, \qquad i, j, k = 1, \ldots, n\,. \tag{1}$$

Die r holonomen Nebenbedingungen werden über $(n - r)$ generalisierte Koordinaten $\bar q_\alpha$ mittels der Transformation

$$q_i = g_i(\bar q_\alpha, t); \qquad i = 1, \ldots, n; \quad \alpha = 1, \ldots, (n - r)\,, \tag{2}$$

berücksichtigt, so daß die Bindungen $f_\varrho = 0$ identisch erfüllt werden. Mit (2) gilt:

$$\dot q_i = \sum_\alpha \frac{\partial g_i}{\partial \bar q_\alpha} \dot{\bar q}_\alpha + \frac{\partial g_i}{\partial t}\,, \qquad \delta q_i = \sum_\alpha \frac{\partial g_i}{\partial \bar q_\alpha} \delta \bar q_\alpha\,. \tag{3a, b}$$

Durch Einsetzen von (2) und (3a, b) in (1) erhält man in bekannter Weise für das Variationsproblem (1)

$$\int_{t_0}^{t_1} dt \left\{\delta\left[\frac{1}{2} \sum_{i,k} a_{ik}(\bar q_\gamma, t) \left(\sum_\alpha \frac{\partial g_i}{\partial \bar q_\alpha} \dot{\bar q}_\alpha + \frac{\partial g_i}{\partial t}\right) \left(\sum_\beta \frac{\partial g_k}{\partial \bar q_\beta} \dot{\bar q}_\beta + \frac{\partial g_k}{\partial t}\right) - \right.\right.$$

$$\left.\left. - V(\bar q_\alpha, t)\right] + \sum_\alpha \sum_i Q_i(\bar q_\gamma, \dot{\bar q}_\gamma, t) \frac{\partial g_i}{\partial \bar q_\alpha} \delta \bar q_\alpha\right\} = 0\,, \tag{4}$$

also die Gestalt

$$\int_{t_0}^{t_1} dt \left\{\delta \bar L(\bar q_\alpha, \dot{\bar q}_\alpha, t) + \sum_\alpha \bar Q_\alpha(\bar q_\gamma, \dot{\bar q}_\gamma, t)\, \delta \bar q_\alpha\right\} = 0;$$

$$\alpha, \gamma = 1, \ldots, (n - r)\,. \tag{5}$$

Das (1) entsprechende HAMILTONsche Variationsproblem lautet

$$\int_{t_0}^{t_1} dt \left\{\delta\left[\sum_i p_i \dot q_i - H(p_i, q_i, t) + \sum_\varrho \lambda_\varrho f_\varrho(q_i, t)\right] + \right.$$

$$\left. + \sum_i Q_i(q_j, p_j, t)\, \delta q_i\right\} = 0 \tag{6a}$$

mit

$$H = \tfrac{1}{2} \sum_{i,k} b_{ik} p_i p_k + V(q_i, t)\,, \tag{6b}$$

sowie

$$\sum_k a_{ik} b_{kj} = \delta_{ij}; \quad \text{d. h.} \quad \boldsymbol{A} \cdot \boldsymbol{B} = \boldsymbol{I}; \quad \boldsymbol{A} = \boldsymbol{A}^T, \quad \boldsymbol{B} = \boldsymbol{B}^T\,. \tag{6c}$$

Um die Transformation (2) zu einer kanonischen zu ergänzen, machen wir für die verallgemeinerte Erzeugende S den Ansatz

$$S = S(p_i, \bar q_\alpha, t) = -\sum_{i=1}^{n} p_i g_i(\bar q_\alpha, t)\,, \tag{7}$$

und erhalten folgende verallgemeinerte kanonische Transformation:

$$q_i = -\frac{\partial S}{\partial p_i} = g_i(\bar q_\alpha, t); \quad \bar p_\alpha = -\frac{\partial S}{\partial \bar q_\alpha} = \sum_i p_i \frac{\partial g_i}{\partial \bar q_\alpha}\,;$$

$$\frac{\partial S}{\partial t} = -\sum_i p_i \frac{\partial g_i}{\partial t}\,. \tag{8a, b, c}$$

Die transformierte HAMILTONfunktion $\bar H$ erhalten wir in zwei Schritten. Wir bilden erstens

$$\bar H' = H + \frac{\partial S}{\partial t} = H(p_i, \bar q_\alpha, t) - \sum_i p_i \frac{\partial g_i(\bar q_\alpha, t)}{\partial t}\,, \tag{9}$$

wobei wir in $\bar H'$ die Transformation $q_i \to \bar q_\alpha$ ausführen. Jedoch ist es nicht möglich, aus den $(n - r)$ Gleichungen (8b) die n Transformationsbeziehungen $p_i \to \bar p_\alpha$ abzuleiten. Daher müssen wir zweitens für $\bar H$ ansetzen

$$\bar H = \bar H' - \sum_{\alpha=1}^{n-r} \bar\lambda_\alpha \left(\sum_i p_i \frac{\partial g_i}{\partial \bar q_\alpha} - \bar p_\alpha\right), \tag{10}$$

womit wir im transformierten Variationsproblem sogar $(n - r)$ Nebenbedingungen mit Multiplikatoren $\bar\lambda_\alpha$ einbauen müssen. Für die $\bar\lambda_\alpha$ werden wir aber algebraische Gleichungen erhalten. Das transformierte HAMILTONsche Variationsproblem lautet somit:

$$\int_{t_0}^{t_1} dt \left\{\delta\left[\sum_\alpha \bar p_\alpha \dot{\bar q}_\alpha - \bar H(p_i, \bar p_\alpha, \bar q_\alpha, \bar\lambda_\alpha, t)\right] + \sum_\alpha \bar Q_\alpha(\bar q_\beta, p_k, t)\, \delta \bar q_\alpha\right\} = 0\,, \tag{11a}$$

$$\bar H = \frac{1}{2} \sum_{i,k} b_{ik}(\bar q_\beta, t) p_i p_k + V(\bar q_\alpha, t) - \sum_i p_i \frac{\partial g_i}{\partial t} -$$

$$- \sum_\alpha \bar\lambda_\alpha \left(\sum_i p_i \frac{\partial g_i}{\partial \bar q_\alpha} - \bar p_\alpha\right), \tag{11b}$$

$$\bar Q_\alpha = \sum_i Q_i(\bar q_\beta, p_k, t) \frac{\partial g_i}{\partial \bar q_\alpha}\,. \tag{11c}$$

Die zugehörigen Bewegungsdifferentialgleichungen sind:

$$\partial \bar p_\alpha: \quad \dot{\bar q}_\alpha = \frac{\partial \bar H}{\partial \bar p_\alpha} = \bar\lambda_\alpha; \qquad \delta \bar q_\alpha: \quad \dot{\bar p}_\alpha = -\frac{\partial \bar H}{\partial \bar q_\alpha} + \bar Q_\alpha\,. \tag{12a, b}$$

Die übrigen Variationen liefern algebraische Gleichungen zur Bestimmung der Variablen p_i und $\bar\lambda_\alpha$.

$$\delta p_i: \quad \frac{\partial \bar H}{\partial p_i} = 0 \to \sum_k b_{ik} p_k - \frac{\partial g_i}{\partial t} - \sum_\alpha \bar\lambda_\alpha \frac{\partial g_i}{\partial \bar q_\alpha} = 0\,, \tag{13a}$$

$$\delta \bar\lambda_\alpha: \quad \frac{\partial \bar H}{\partial \bar\lambda_\alpha} = 0 \to \sum_i p_i \frac{\partial g_i}{\partial \bar q_\alpha} - \bar p_\alpha = 0\,. \tag{13b}$$

Aus (13a) folgt unter Beachtung von (6c):

$$p_s = \sum_i a_{si} \frac{\partial g_i}{\partial t} + \sum_i \sum_\alpha \bar\lambda_\alpha a_{si} \frac{\partial g_i}{\partial \bar q_\alpha}\,. \tag{14}$$

Damit kann man in (13b) die n alten Impulse p_s eliminieren:

$$\bar p_\beta = \sum_s p_s \frac{\partial g_s}{\partial \bar q_\beta} = \sum_{i,s} a_{si} \frac{\partial g_i}{\partial t} \frac{\partial g_s}{\partial \bar q_\beta} + \sum_{i,s} \sum_\alpha \bar\lambda_\alpha a_{si} \frac{\partial g_i}{\partial \bar q_\beta} \frac{\partial g_s}{\partial \bar q_\beta}\,. \tag{15}$$

und hat mit (15) ein lineares Gleichungssystem zur Bestimmung der $(n - r)$ Multiplikatoren $\bar\lambda_\gamma = \bar\lambda_\gamma(\bar p_\delta, \bar q_\delta, t)$, welche linear in den $\bar p_\delta$ sind, gefunden.

Setzt man die $\bar\lambda_\alpha$ und p_i in die rechten Seiten von (12a, b) ein, so hängen dann diese nur noch von den generalisierten Variablen $\bar q_\alpha$ und $\bar p_\alpha$ sowie der Zeit t ab, d. h. die r holonomen Nebenbedingungen sind berücksichtigt. Zu denselben Differentialgleichungen in $\bar q_\alpha$, $\bar p_\alpha$ und t gelangt man, wenn man die Größen p_i und $\bar\lambda_\alpha$ direkt im Variationsproblem (11a, b, c) durch die neuen Variablen $\bar q_\alpha$ und $\bar p_\alpha$ ersetzt. Man erhält dann direkt das klassische HAMILTONsche Variationsproblem in den neuen Variablen $\bar q_\alpha$, $\bar p_\alpha$:

$$\int_{t_0}^{t_1} dt \left\{\delta\left[\sum_\alpha \bar p_\alpha \dot{\bar q}_\alpha - \bar H(\bar p_\alpha, \bar q_\alpha, t)\right] + \sum_\alpha \bar Q_\alpha(\bar p_\beta, \bar q_\beta, t)\, \delta \bar q_\alpha\right\} = 0\,. \tag{16}$$

Wir wollen nun die Äquivalenz zwischen dem HAMILTONformalismus, gegeben durch (11a, b, c) unter Benutzung von (12), (13), (14), (15), und dem LAGRANGEformalismus, gegeben durch (4), zeigen. Dazu müssen wir die p_i, $\bar p_\alpha$, $\bar\lambda_\alpha$ durch $\bar q_\alpha$, $\dot{\bar q}_\alpha$ und t ausdrücken. Wir benutzen im folgenden die Vektorschreibweise. Mit (13b), also $\bar{\boldsymbol{p}} = \dfrac{\partial \boldsymbol{g}^T}{\partial \bar{\boldsymbol{q}}} \boldsymbol{p}$, eliminieren wir die Variable $\bar{\boldsymbol{p}}$, und erhalten mit $\bar{\boldsymbol{\lambda}} = \dot{\bar{\boldsymbol{q}}}$ (12a) für (14):

$$\boldsymbol{p} = \boldsymbol{A}\left\{\left(\frac{\partial \boldsymbol{g}^T}{\partial \bar{\boldsymbol{q}}}\right)^T \dot{\bar{\boldsymbol{q}}} + \frac{\partial \boldsymbol{g}}{\partial t}\right\}. \tag{17}$$

Damit kann man zeigen, daß mit (6c) gilt:

$$\bar{\boldsymbol{p}}^T \dot{\bar{\boldsymbol{q}}} + \boldsymbol{p}^T \frac{\partial \boldsymbol{g}}{\partial t} = \boldsymbol{p}^T \left[\left(\frac{\partial \boldsymbol{g}^T}{\partial \bar{\boldsymbol{q}}}\right)^T \dot{\bar{\boldsymbol{q}}} + \frac{\partial \boldsymbol{g}}{\partial t}\right]$$

$$= \left[\left(\frac{\partial \boldsymbol{g}^T}{\partial \bar{\boldsymbol{q}}}\right)^T \dot{\bar{\boldsymbol{q}}} + \frac{\partial \boldsymbol{g}}{\partial t}\right]^T \boldsymbol{A} \left[\left(\frac{\partial \boldsymbol{g}^T}{\partial \bar{\boldsymbol{q}}}\right)^T \dot{\bar{\boldsymbol{q}}} + \frac{\partial \boldsymbol{g}}{\partial t}\right] = \boldsymbol{p}^T \boldsymbol{B} \boldsymbol{p}\,. \tag{18}$$

Das Ausdrücken der Variablen $\boldsymbol{p}$, $\boldsymbol{q}$, $\bar{\boldsymbol{\lambda}}$ durch $\bar{\boldsymbol{q}}$, $\dot{\bar{\boldsymbol{q}}}$, t in (11a, b, c) führt also zum Variationsproblem

$$\int_{t_0}^{t_1} dt \left\{\delta\left[\frac{1}{2}\left(\left(\frac{\partial \boldsymbol{g}^T}{\partial \bar{\boldsymbol{q}}}\right)^T \dot{\bar{\boldsymbol{q}}} + \frac{\partial \boldsymbol{g}}{\partial t}\right)^T \boldsymbol{A} \left(\left(\frac{\partial \boldsymbol{g}^T}{\partial \bar{\boldsymbol{q}}}\right)^T \dot{\bar{\boldsymbol{q}}} + \frac{\partial \boldsymbol{g}}{\partial t}\right) + \right.\right.$$

$$\left. + Q^T \left(\frac{\partial \boldsymbol{g}^T}{\partial \bar{\boldsymbol{q}}}\right)^T \delta \bar{\boldsymbol{q}}\right\} = 0\,, \tag{19}$$

welches mit (4) identisch ist.

Als **Beispiel** betrachten wir einen Massenpunkt im Schwerefeld (x_1: Richtung der Schwerkraft), dessen Bewegung auf einem

Kreis mit dem Radius $a = a(t)$ erzwungen wird bei geschwindigkeitsproportionaler Dämpfung. Das Variationsproblem entsprechend (6) lautet:

$$\int_{t_0}^{t_1} \mathrm{d}t \left\{\delta \left[p_1\dot{x}_1 + p_2\dot{x}_2 - \left\{\frac{1}{2m}(p_1^2 + p_2^2) - mgx_1\right\} + \lambda(x_1^2 + x_2^2 - a^2(t))\right] - d(p_1\delta x_1 + p_2\delta x_2)\right\} = 0 . \quad (20)$$

Für die Erzeugende S setzen wir an

$$S = S(p_1, p_2, \varphi, t) = -(p_1 a \cos\varphi + p_2 a \sin\varphi) \quad (21)$$

und erhalten die Transformationsgleichungen

$$x_1 = a\cos\varphi; \quad x_2 = a\sin\varphi; \quad p = a(-p_1 \sin\varphi + p_2\cos\varphi)\,,$$
$$\frac{\partial S}{\partial t} = -(p_1\cos\varphi + p_2\sin\varphi)\,\dot{a}\,. \quad (22)$$

Analog (11) erhalten wir:

$$\int_{t_0}^{t_1} \mathrm{d}t \left\{\delta \left[p_\varphi\dot{\varphi} - \left\{\frac{1}{2m}(p_1^2 + p_2^2) - mga\cos\varphi - (p_1\cos\varphi + p_2\sin\varphi)\,\dot{a} - \bar{\lambda}(-p_1 a\sin\varphi + p_2 a\cos\varphi - p_\varphi)\right\}\right] - \mathrm{d}p_\varphi\,\delta\varphi\right\} = 0 . \quad (23)$$

Die Variation liefert

$$\delta p\colon \quad \dot{\varphi} = \bar{\lambda}\,, \quad (24\text{a})$$

$$\delta\varphi\colon \quad \dot{p}_\varphi = -mga\sin\varphi + \dot{a}(-p_1\sin\varphi + p_2\cos\varphi) + \bar{\lambda}a(-p_1\cos\varphi - p_2\sin\varphi) - \mathrm{d}p_\varphi \quad (24\text{b})$$

$$\delta p_1\colon \quad p_1 = m(\dot{a}\cos\varphi - \bar{\lambda}a\sin\varphi)\,, \quad (24\text{c})$$

$$\delta p_2\colon \quad p_2 = m(\dot{a}\sin\varphi + \bar{\lambda}a\cos\varphi)\,, \quad (24\text{d})$$

$$\delta\bar{\lambda}\colon \quad a(-p_1\sin\varphi + p_2\cos\varphi) - p_\varphi = 0\,. \quad (24\text{e})$$

Drückt man über (24c, d, e) die Größen p_1, p_2, $\bar{\lambda}$ durch φ, p_φ und t aus und setzt dies in (20) ein, so gelangt man direkt zum Variationsproblem in den generalisierten Variablen φ und p_φ:

$$\int_{t_0}^{t_1} \mathrm{d}t \left\{\delta \left[p_\varphi\dot{\varphi} - \left\{\frac{p_\varphi^2}{2ma^2} - \frac{m\dot{a}^2}{2} - mga\cos\varphi\right\}\right] - \mathrm{d}p_\varphi\,\delta\varphi\right\} = 0\,. \quad (25)$$

Der Verfasser dankt der Deutschen Forschungsgemeinschaft für ihre Unterstützung.

Literatur

1 VOLZ, HELMUT, Einführung in die Theoretische Mechanik II, Akademische Verlagsgesellschaft Frankfurt am Main (1972), S. 48.

Eingereicht am 5. 5. 1980 (Kurzfassung eines Vortrages auf der GAMM-Tagung 1980 in Berlin (West))

Anschrift: Prof. Dr.-Ing. J. BAUMGARTE, Mechanik-Zentrum, Technische Universität Braunschweig, D-3300 Braunschweig, BRD

BUCHBESPRECHUNGEN

McGuire, W. / Gallagher, R. H., Matrix Structural Analysis. New York-Chichester-Brisbane-Toronto, John Wiley & Sons 1979. XIX, 460 S., zahllose Abb., £ 12.50. ISBN 0-471-03059-7

Gegenstand des Buches ist die Statik der Stabtragwerke für Studenten, die zu Beginn des Fachstudiums eine entsprechende Vorlesung besuchen. Neben Verformungs- und Kraftgrößenmethode werden zur Berechnung statisch unbestimmter Systeme auch Elementansätze mit HERMITEschen Interpolationspolynomen, ausgehend vom Prinzip der virtuellen Arbeiten, angewendet. Behandelt werden hauptsächlich Rahmen und Stockwerkrahmen. Die Methode der finiten Elemente für Flächentragwerke wird im letzten Kapitel kurz und allgemein beschrieben; damit soll sicher nur ein Ausblick auf einen nachfolgenden Kurs gegeben werden. Ein umfangreicheres Kapitel über Lösungsmethoden für lineare Gleichungssysteme enthält auch Fehlerbetrachtungen.

Die Besonderheit des Buches ist die ungewöhnlich große Anzahl von gelösten Beispielen, die graphisch sehr übersichtlich angeordnet sind. Die Kombination von Leitfaden und Aufgabensammlung ist im vorliegenden wie auch in anderen amerikanischen Lehrbüchern vor allem durch sorgfältige drucktechnische Gestaltung vorbildlich.

Berlin L. KNÖFEL

Scholz, Ch., Handbuch der Magnetbandspeichertechnik. Berlin, VEB Verlag Technik 1979. 392 S., 415 Abb., M 30,–. BN 552 626 8

Das vorliegende Buch zeichnet sich durch die Fülle des zusammengestellten Materials aus dem gesamten Spektrum der Magnetspeichertechnik aus. Man findet Angaben zu Einsatzmöglichkeiten magnetischer Speicher; ein Kapitel ist den Verfahren zur Speicherung analoger bzw. digitaler Signale gewidmet. Weitere Kapitel behandeln die Magnetbänder, die Magnetköpfe (inclusive deren Anschlußschaltungen) sowie die Transportwerke. So erweist sich die gewählte Tabellenform der Darstellung oft als die für ein Handbuch einzig mögliche Form, des Inhaltes Herr zu werden. Jedoch darf das nicht auf Kosten der Verständlichkeit gehen; insbesondere der Anhang („9. Grundlagen zur Technik der Aufzeichnung und Wiedergabe von Signalen") leidet unter der Kürze. Gerade diesen hätte man sich ausführlicher und aktueller gewünscht, zumal sich am Anfang der 70er Jahre auf dem Gebiet der Hysterese- und Aufzeichnungsvorgangs-Modelle einiges getan hat.

Nichtsdestotrotz dürfte es ein nützliches Nachschlagewerk für die mit der Materie vertrauten Techniker sein.

Berlin H. STRESE

Prékopa, A. (Ed.), Studies on Mathematical Programming. Mathematical Methods of Operations Research 1. Budapest, Akadémiai Kiadó 1980. 200 S., ISBN 963-05-1854-6

Der vorliegende Band enthält die zum Teil erweiterten Vorträge, die im Rahmen der „Third Conference on Mathematical Programming" (31. 1.–6. 2. 1975 in Mátrafüred, Ungarn) gehalten wurden. Im einzelnen werden die folgenden Themen behandelt: D. E. BELL gibt in seiner Arbeit „The group approach to integer programming: an outline of recent progress" einen kurzen Überblick über die Entwicklung der von GOMORY stammenden Gruppen diskreter Optimierungsprobleme. In der Arbeit „On the theory of conjugate functions" von R. DEUMLICH und K.-H. ELSTER wird eine Verallgemeinerung der konjugierten Funktion untersucht. J. DUPAĆOVÁ faßt in ihrer Arbeit „On Minimax Decision rule in stochastic linear programming" Ergebnisse ihrer Untersuchungen zusammen über Minimax-Entscheidungen bei stochastischen Optimierungsproblemen, bei denen die Verteilungsfunktion unbekannt ist. I. I. YEREMIN und V. D. MAZUROV geben in ihrer Arbeit „Synthesis of nonstationary processes of mathematical programming, containing pattern recognition" einen knappen Abriss über den Entwurf eines Prozesses für die Lösung eines speziellen Optimierungsproblemes und diskutieren verschiedene Anwendungen. In seiner Arbeit „On the qualitative stability in the nonlinear parametric programming and some applications" berichtet J. Guddat u. a. über Stabilitätssätze für konvexe quadratische Optimierungsaufgaben sowie über den Zusammenhang zwischen parametrischer Optimierung und Polyoptimierung. H. Hollatz beschreibt in seiner Arbeit „An analytic geometric approach to necessary conditions for minimax problems", wie man in einheitlicher Weise notwendige Bedingungen für Minimaxprobleme herleiten kann. In der Arbeit „Synthése et agrégation des méthodes en programmation mathématique" faßt P. Huard Ergebnisse aus dem Gebiet der Optimierung zusammen, die er und seine Kollegen an der Universität Lille in den letzten Jahren erzielt haben. In seiner einführenden Arbeit „Introduction to isotonic estimation" untersucht W. S. Jewell beste Mengen von Schätzfunktionen über endlichen Klassen, wenn zwischen den Schätzfunktionen verschiedener Klassen Ungleichungen bestehen. G. Kéri beschreibt eine Methode zur Inversion von Matrizen. L. Lovász gibt in seiner Arbeit „Efficient algorithms: an approach by formal logic" eine neue Formulierung der Berechnungskomplexität, die auf formaler Logik beruht. In der Arbeit „Dynamic type stochastic programming models" gibt A. Prékopa einen Überblick über gewisse Modelle stochastischer Programme, die bei Anwendungen von Interesse sind. H. Weinert behandelt in seiner Arbeit „On the solution of certain multiparametric linear programming problems" parametrische lineare Optimierungsaufgaben, bei denen Parameter sowohl in der Zielfunktion als auch im Restriktionsvektor auftreten können. In der Arbeit „Basic factorization for block-angular linear programs" untersucht H. Winkler lineare Programme mit Blockstruktur und beschreibt Algorithmen zu deren Lösung. K. Zimmermann verwendet in seiner Arbeit „Solution of some optimization problems on the extremal algebra" den Begriff der extremalen Algebra zur Lösung gewisser Optimierungsprobleme. Schließlich entwickeln S. Zionts und J. Wallenius in ihrer Arbeit „On finding the subset of efficient vectors of an arbitrary set of vectors" eine Methode, um die effizienten Vektoren bei Vektoroptimierungsproblemen zu finden.

Frankfurt B. Brosowski

Großmann, C. / Kaplan, A. A., Strafmethoden und modifizierte Lagrangefunktionen in der nichtlinearen Optimierung. Teubner-Texte zur Mathematik. Leipzig, BSB B. G. Teubner Verlagsgesellschaft 1979. 200 S., 10 Abb., M 18,–. BN 6659177

In vorliegendem Buch werden einige Straf-, Zentren- und verallgemeinerte Lagrangemethoden zur Lösung finiter nichtlinearer Optimierungsaufgaben dargestellt. Das Wesen dieser Verfahren besteht darin, daß Lösungen von Aufgaben mit Nebenbedingungen mit Hilfe von Lösungen geeigneter Folgen unrestringierter Aufgaben approximiert werden.

Die Autoren stellen in leicht verständlicher Weise zunächst die benötigten theoretischen Grundlagen der nichtlinearen Optimierung und eine Theorie der betrachteten Verfahrensklassen dar. Untersuchungen zur Konvergenzgeschwindigkeit und zu einigen Fragen, die bei der numerischen Realisierung der Verfahren auftreten, wird ausreichend Raum gewährt, so daß das Buch nicht nur als Einführung für den Studierenden, sondern auch dem Anwender, der in der Praxis nichtlineare Optimierungsaufgaben zu bearbeiten hat, empfohlen werden kann. Es werden keine Vorkenntnisse aus der Optimierung vorausgesetzt. Mit wenigen Ausnahmen ist die wichtigste Literatur bis etwa 1977 berücksichtigt.

Darmstadt R. Horst

Steiner, H.-G. (Hrsg.), Didaktik der Mathematik. Wege der Forschung CCCLXI. Darmstadt, Wissenschaftliche Buchgesellschaft 1978. XLVIII, 478 S., ISBN 3-534-06005-9

In der DDR beziehbar im Rahmen des Kontingents.

Dies ist eine Sammlung von zwanzig sehr verschiedenartigen, schon früher und anderweitig veröffentlichten Aufsätzen, die — von zwei älteren abgesehen — der Dekade von 1964 bis 1974 entstammen, die umfangreiche Bibliographie reicht bis 1977. Die Arbeiten gliedern sich in drei Themenkreise: I. Zur Orts- und Aufgabenbestimmung der Mathematikdidaktik; II. Empirische Forschung — Psychologische und lerntheoretische Aspekte; III. Gegenstandsdidaktische Analyse und Strukturierung.

Besonders hervorzuheben, weil von allgemeinerem Interesse, wären vielleicht die Beiträge zur Didaktik der Wahrscheinlichkeitsrechnung, für deren Einführung in den Schulunterricht offenbar die Zeit heranreift; die Situation hat hier eine gewisse Ähnlichkeit mit der vor hundert Jahren, als es um die große Revolution der Einführung der Infinitesimalrechnung ging; kleine Umstürze, wie z. B. die Abschaffung der sphärischen Trigonometrie oder die Aufnahme der Vektorrechnung, gibt es ja öfter mal. — Ein weiterer Aufsatz des Herausgebers, bescheiden nur mit „Einleitung" betitelt, gibt einen Überblick über die historische Entwicklung der Mathematikdidaktik, vornehmlich in Deutschland. — Geradezu wohltuend ist es, daß bei der Auswahl der große Renner der letzten Jahre, die „Mengenlehre", keine Rolle spielte, ist man sich doch noch immer nicht einig, ob diese Idee enorm oder abnorm war.

Sammelbände sind im allgemeinen inhomogen, will sagen, sie können oft nur ein lückenhaftes und unorganisches Bild des jeweiligen Gesamtgebietes geben. Im Falle der Didaktik der Mathematik wäre ein solches nun genau das richtige Bild, jedenfalls wenn man sich folgendem Zitat (S. 271) anschließt: „Da die mathematikdidaktischen Untersuchungen selten als Teil eines aufeinander abgestimmten Forschungsprogramms durchgeführt werden, da sie selten an einer Theorie aufgehängt sind, es sei denn mit dem dünnsten aller Fäden, und da es typisch für sie ist, daß sie von Leuten durchgeführt werden, die vorher nicht wissenschaftlich gearbeitet haben und es wahrscheinlich auch nachher nie wieder tun werden, überrascht es nicht, daß sie sich nicht organisch zusammenfügen". Trotz dieses Standes wird man aber immerhin sagen dürfen, daß sich didaktische Fragen der Mathematik heute mit der Ansicht von Goethes Faust — „es trägt Verstand und rechter Sinn mit wenig Kunst sich selber vor" — nicht mehr erledigen lassen. Auch das zeigt dieser Band.

Berlin W. Heinrich

Browne, B. T. / Miller, J. J. H., Numerical Analysis of Semiconductor Devices. Proceedings of the NASECODE I Conference held at Trinity College, Dublin, June, 1979. Dublin, Boole Press 1979. XI, 303 S., £ 20.00. ISBN 0-906783-00-3

Für den Entwurf mikroelektronischer Schaltkreise ist es unter anderem erforderlich, das physikalische Verhalten von Transistoren in mathematischen Gleichungen darzustellen und diese numerisch zu lösen. Hauptanliegen der Konferenz war die numerische Behandlung von Gleichungen für Halbleiterbauelemente. In mehreren Vorträgen wurden Verfahren zur numerischen Lösung der Gleichungen für Transistoren, ein System bestehend aus Poissongleichung und Transportgleichungen, vorgestellt, über die numerischen Erfahrungen berichtet und die Ergebnisse physikalisch diskutiert. Hierzu werden die Poissongleichung vielfach durch finite Elemente und die Transportgleichungen durch Differenzenverfahren approximiert und die entstehenden linearen Gleichungssysteme durch verschiedene Methoden, u. a. mit speziellen Sparse-Matrix Techniken, behandelt. — In weiteren Vorträgen werden Programmpakete für den Entwurf von MOS-Transistoren und deren Analysis, effektive 3-dimensionale Bipolar Transistor Analyse-Programme, 2-dimensionale Analysis von I^2L-Elemente u. a. vorgestellt. Außerdem wird auf andere wichtige Aspekte der numerischen Berechnung von Halbleitern eingegangen. — Das Buch enthält die Manuskripte von neun eingeladenen und die Kurzvorträge von 26 weiteren Vortragenden.

Berlin F. Grund

EINGEGANGENE BÜCHER

Bei der Schriftleitung sind folgende Bücher eingegangen (ausführliche Besprechung bleibt vorbehalten).

Löthe, H. / Müller, K. P., Taschenrechner. Erläuterungen zum Aufbau und Vorschläge für den Unterricht. Stuttgart, B. G. Teubner 1979. 168 S., 49 Abb., DM 18,80. ISBN 3-519-02711-9

Pham, F., Singularités des systèmes différentiels de Gauss-Manin. Progress in Mathematics 2. Boston-Basel-Stuttgart, Birkhäuser Verlag 1979. 339 S., sFr. 34.—. DM 37,—. $ 16.—. ISBN 3-7643-3002-3

Elliott, P. D. T. A., Probabilistic Number Theory. I: Mean-Value Theorems. XXII, 393 S., DM 70,—. US $ 38.50. 1979. II: Central Limit Theorems. XVIII, 375 S., DM 70,—. US $ 38.50. Berlin-Heidelberg-New York, Springer-Verlag. 1980. Grundlehren der mathematischen Wissenschaften 239, 240.

Dynkin, E. B. / Yushkevich, A. A., Controlled Markov Processes. Grundlehren der mathematischen Wissenschaften 235. Berlin-Heidelberg-New York, Springer-Verlag 1979. XVII, 289 S., 36 Abb., DM 79,50. US $ 46.90. ISBN 3-540-90387-9

Lehrbuch Höhere Festigkeitslehre. Band 1: Grundlagen der Höheren Festigkeitslehre. Autorenkollektiv unter Federführung von Göldner, H., Leipzig, VEB Fachbuchverlag 1979. 236 S., 116 Abb., M 16,—. Auslandspreis M 24,—. BN 546 434 3

Sniatycki, J., Geometric Quantization and Quantum Mechanics. Applied Mathematical Sciences 30. Berlin-Heidelberg-New York, Springer-Verlag 1980. IX, 230 S., DM 28,—. US $ 16.60. ISBN 3-540-90469-7

Glaeser, G. (Hrsg.), Didaktik mathematischer Probleme und Aufgaben. Braunschweig, Vieweg 1980. X, 214 S., 162 Abb., DM 29,80. ISBN 3-528-08395-6

Haken, H. (Ed.), Dynamics of Synergetic Systems. Proceedings of the International Symposium on Synergetics. Bielefeld, September, 1979. Springer Series in Synergetic 6. Belin-Heidelberg-New York, Springer-Verlag 1980. VIII, 271 S., 146 Abb., DM 68,—. US $ 40.20. ISBN 3-540-09918-2

Whitham, G. B., Lectures on Wave Propagation. Berlin-Heidelberg-New York, Springer-Verlag 1979. Bombay Tata Institute of Fundamental Research 1979. VII, 148 S., 31 Abb., DM 18,—. US $ 8.00. ISBN 3-540-08945-4

The Monte Carlo Methods in Atmospheric Optics. Marchuk, G. I. / Mikhailov, G. A. / Nazaraliev, M. A. / Darbinjan, R. A. / Kargin, B. A. / Elepov, B. S., Springer Series in Optical Sciences 12. Berlin-Heidelberg-New York, Springer-Verlag 1980. VIII, 208 S., 44 Abb., DM 59,—. US $ 34.90. ISBN 3-540-09402-4

Herschel, R. / Pieper, F., PASCAL, Systematische Darstellung von PASCAL und CONCURRENT PASCAL für den Anwender. Reihe Datenverarbeitung. München-Wien, R. Oldenbourg. 1979. 248 S., 183 Abb., DM 23,80. ISBN 3-486-23681-4

Wells, R. O., Differential Analysis on Complex Manifolds, 2nd ed., Graduate Texts in Mathematics 65. Berlin-Heidelberg-New York, Springer-Verlag 1980. X, 260 S., DM 39,50. US $ 23.30. ISBN 3-540-90419-0

Plachky, D. / Baringhaus, L. / Schmitz, N., Stochastik I, Eine elementare Einführung in Grundbegriffe der Wahrscheinlichkeitsrechnung und Statistik. Studienbuch für Studierende der Mathematik, Natur- und Wirtschaftswissenschaften ab 2. Semester. Wiesbaden, Akademische Verlagsgesellschaft 1978. 250 S., DM 22,—. ISBN 3-400-00390-5

Oden, J. T. (Ed.), Computational Methods in Nonlinear Mechanics. Proceedings of the TICOM Second Intern. Conference, Austin, Texas, March 1979. Amsterdam-New York, North-Holland Publ. Co. 1980. VIII, 540 S., US $ 68.25. Dfl. 140.00. ISBN 0-444-85382-0

Akkaş, N. (Ed.), Progress in Biomechanics. Proceedings of the NATO Advanced Study Institute, Ankara, July 1978. Series E: Applied Science 32. Alphen aan den Rijn, Sijthoff & Noordhoff 1979. 395 S., Dfl. 75.00. $ 37.50. ISBN 90-286-0479-0

Pekalski, A. / Przystawa, J. (Eds.), Modern Trends in the Theory of Condensed Matter. Proceedings, Karpacz, 1979. Lecture Notes in Physics 115. Berlin-Heidelberg-New York, Springer-Verlag 1980. IX, 597 S., DM 68,—. US $ 40.20. ISBN 3-540-09752-X

Maess, G., Iterative Lösung linearer Gleichungssysteme. Nova Acta Leopoldina NF 238/Bd. 52. Halle, Deutsche Akademie der Naturforscher Leopoldina 1979. 79 S., M 24,—.

Hudson, J. A., The Excitation and Propagation of Elastic Waves. Cambridge Monographs on Mechanics and Applied Mathematics. Cambridge, Cambridge University Press 1980. VIII, 226 S., £ 14.00 A H/C. ISBN 0-521-22777-1

Hirschel, E. H. (Ed.), Proceedings of the Third GAMM-Conference on Numerical Methods in Fluid Mechanics. Notes on Numerical Fluid Mechanics 2. DVFLR, Köln, October 1979. Braunschweig, Vieweg 1980. VIII, 315 S., DM 98,—. ISBN 3-528-08076-0

Karl, H. (Hrsg.), Datennetze, 1. Treffen 1978 des German Chapter of the ACM am 2. 6. 1978 in Wiesbaden. Applied Computer Science — Berichte zur praktischen Informatik 15. München-Wien, Carl Hanser Verlag 1979. 200 S., 70 Abb., DM 38,—. ISBN 3-446-13016-0

Gray, D. F. / Linsky, J. L. (Eds.), Stellar Turbulence, Proceedings of Colloquium 51 of the International Astronomical Union, London (Canada) 1979. Lecture Notes in Physics 114. Berlin-Heidelberg-New York, Springer-Verlag 1980. IX, 308 S., 97 Abb., DM 37,60. US $ 22.20. ISBN 3-540-09737-6

Decuyper, M., Exercices de mécanique eulérienne. Paris-New York-Barcelone-Milan, Masson 1980. 351 S., ISBN 2-225-65351-8

Witting, H., Mathematische Statistik. Teubner Studienbücher Mathematik, LAMM 9. Eine Einführung in Theorie und Methoden. Stuttgart, B. G. Teubner 1978. 223 S., DM 26,80. ISBN 3-519-22036-9

Mardia, K. V. / Kent, J. T. / Bibby, J. M., Multivariate Analysis. London-New York-Toronto-Sydney-San Francisco, Academic Press 1979. XV, 521 S., $ 34.00 P/B., ISBN 0-12-471252-5

Alefeld, G. / Grigorieff, R. D. (Eds.), Fundamentals of Numerical Computation (Computer-Oriented Numerical Analysis). Edited in Cooperation with Albrecht, R. / Kulisch, U. / Stummel, F., Computing Supplementum 2. Wien-New York, Springer-Verlag 1980. VII, 229 S., 34 Abb., DM 98,—. S 690,—. US $ 57.90. ISBN 3-211-81566-X

Parlett, B. N., The Symmetric Eigenvalue Problem. Prentice-Hall Series in Computational Mathematics. Englewood Cliffs, N.J. 07632, Prentice-Hall, Inc. 1980. XX, 348 S., $ 25.00. ISBN 0-13-880047-2

Cottle, R. W. / Giannessi, F. / Lions, J.-L., Variational Inequality and Complementarity Problems. Theory and Applications. Chichester-New York-Brisbane-Toronto, John Wiley & Sons 1980. XVII, 408 S., £ 17.75. ISBN 0-471-27610-3

Winogradzki, J., Calcul tensoriel dans un continuum amorphe. Les méthodes tensorielles de la physique 1. Paris-New York-Barcelone-Milan, MASSON 1979. VIII, 219 S., ISBN 2-225-49269-7

DFVLR-Forschungsbereich Energetik. Wissenschaftlicher Bericht (Stand 1980) des Forschungsbereichs Energetik der DFVLR. Köln. 107 S., zahlreiche Abb., DM 20,—.

Codes for Boundary-Value Problems in Ordinary Differential Equations. Proceedings of a Working Conference, May, 1978. Edited by Childs, B. / Scott, M. / Daniel, J. W. / Denman, E. / Nelson, P., Lecture Notes in Computer Science 76. Berlin-Heidelberg-New York, Springer-Verlag 1979. VIII, 388 S., DM 35,50. US $ 19.60. ISBN 3-540-09554-3

Newell, G. F., Approximate Behavior of Tandem Queues. Lecture Notes in Economics and Mathematical Systems 171. Berlin-Heidelberg-New York, Springer-Verlag 1979. XI, 410 S., 44 Abb., DM 39,—. US $ 21.50. ISBN 3-540-09552-7

Wuytack, L. (Ed.), Padé Approximation and Its Applications. Proceedings, Antwerp, Belgium, 1979. Lecture Notes in Mathematics 765. Berlin-Heidelberg-New York, Springer-Verlag 1979. VI, 392 S., 14 Abb., DM 35,50. US $ 19.90. ISBN 3-540-09717-1

Pacault, A. / Vidal, C. (Eds.), Synergetics far from Equilibrium. Proceedings, Bordeaux, France, 1978. Springer Series in Synergetics 3. Berlin-Heidelberg-New York, Springer-Verlag 1979. IX, 175 S., 109 Abb., DM 59,—. US $ 32.50. ISBN 3-540-09304-4

Schumann, W. / Dubas, M., Holographic Interferometry. From the Scope of Deformation Analysis of Opaque Bodies. Springer Series in Optical Science 16. Berlin-Heidelberg-New York, Springer-Verlag 1979. X, 194 S., 73 Abb., DM 59,—. US $ 32.50. ISBN 3-540-09371-0

von Bally, G. (Ed.), Holography in Medicine and Biology. Proceedings of the International Workshop, Münster, 1979. Springer Series in Optical Sciences 18. Berlin-Heidelberg-New York, Springer-Verlag 1979. IX, 269 S., 224 Abb., DM 78,—. US $ 42.90. ISBN 3-540-09793-7

Dokumentation Rheologie. 1979, Nr. 45. Berlin, Bundesanstalt für Materialprüfung 1979. XV, 848 S., DM 100,—. (for Members of the DRG 80,—). ISBN 0340-8388

Girault, V. / Raviart, P.-A., Finite Element Approximation of the Navier-Stokes Equations. Lecture Notes in Mathematics 749. Berlin-Heidelberg-New York, Springer-Verlag 1979. VII, 200 S., 4 Abb., DM 25,—. US $ 13.80. ISBN 3-540-09557-8

Gasser, Th. / Rosenblatt, M. (Eds.), Smoothing Techniques for Curve Estimation. Proceedings, Heidelberg 1979. Lecture Notes in Mathematics 757. Berlin-Heidelberg-New York, Springer-Verlag 1979. V, 245 S., 29 Abb., DM 25,—. US $ 13.80. ISBN 3-540-09706-6

Sattinger, D. H., Group Theoretic Methods in Bifurcation Theory. Lecture Notes in Mathematics 762. Berlin-Heidelberg-New York, Springer-Verlag 1979. V, 241 S., 14 Abb., DM 25,—. US $ 14.00. ISBN 3-540-09715-5

Leipholz, H. H. E. (Ed.), Structural Control. Proceedings of IUTAM Symposium, Waterloo, Canada, 1979. Amsterdam-New York, North-Holland Publ. Co. 1980. 822 S., US $ 87.75. Dfl. 180.00. ISBN 0-444-85485-1

McIntosh, J. E. A. / McIntosh, R. P., Mathematical Modelling and Computers in Endocrinology. Monographs on Endocrinology 16. Berlin-Heidelberg-New York, Springer-Verlag 1980. XII, 337 S., 73 Abb., DM 88,—. US $ 52.00. ISBN 3-540-09693-0

Anderssen, R. S. / De Hoog, F. R. / Lukas, M. A. (Eds.), The Application and Numerical Solution of Integral Equations. Proceedings. Series Mechanics: Analysis 6. Alphen aan den Rijn, Sijthoff & Noordhoff 1980. 265 S., Dfl. 55.00. $ 27.50. ISBN 90-286-0450-2

Beckenbach, E. F. (Ed.), General Inequalities 2. Proceedings of the Second Intern. Conference, Oberwolfach 1978. ISNM 47. Basel-Boston-Stuttgart, Birkhäuser Verlag 1980. XX, 485 S., sFr. 60,—. ISBN 3-7643-1056-1

Todd, J., Basic Numerical Mathematics, Vol. 1: Numerical Analysis. ISNM 14. Basel-Stuttgart, Birkhäuser Verlag 1979. 253 S., sFr. 54,—. ISBN 3-7643-0729-3

Collatz, L. / Meinardus, G. / Werner, H. (Eds.), Numerical Methods of Approximation Theory. Vortragsauszüge der Tagung, Oberwolfach, März 1979. ISNM 52. Basel-Boston-Stuttgart, Birkhäuser Verlag 1980. 337 S., sFr. 45,—. ISBN 3-7643-1103-7

Kosniowski, C., A First Course in Algebraic Topology. Cambridge, Cambridge University Press 1980. VIII, 269 S., £ 18.00 A H/C (SBN 0-521-23195-7). £ 6.96 A P/B (SBN 0-521-29864-4)

Frehse, J. / Pallaschke, D. / Trottenberg, U. (Eds.), Special Topics of Applied Mathematics. Functional Analysis, Numerical Analysis and Optimization. Proceedings of the Seminar in Bonn, 8—10 October, 1979. Amsterdam-New York, North-Holland Publ. Co. 1980. VIII, 242 S., US $ 39.00. Dfl. 80.00. ISBN 0-444-86035-5

Hestenes, M., Conjugate Direction Methods in Optimization. Applications of Mathematics 12. Berlin-Heidelberg-New York, Springer-Verlag 1980. X, 325 S., 22 Abb., DM 59,50. US $ 35.10. ISBN 3-540-90455-7

Hida, T., Brownian Motion. Applications of Mathematics 11. Berlin-Heidelberg-New York, Springer-Verlag 1980. XVI, 325 S., 13 Abb., DM 68,—. US $ 37.40. ISBN 3-540-90439-5

Kazmierczak, H. (Hrsg.), Erfassung und maschinelle Verarbeitung von Bilddaten. Grundlagen und Anwendungen. Wien-New York, Springer-Verlag 1980. X, 399 S., 236 Abb., DM 168,—. S 1200,—. US $ 99.20. ISBN 3-211-81555-4

Haken, H. / Wolf, H. C., Atom- und Quantenphysik. Eine Einführung in die experimentellen und theoretischen Grundlagen. Berlin-Heidelberg-New York, Springer-Verlag 1980. XIV, 365 S., 228 Abb., DM 54,—. US $ 31.90. ISBN 3-540-09889-5

Brandt, S. / Dahmen, H. D., Physik. Eine Einführung in Experiment und Theorie. Bd. 2: Elektrodynamik. Berlin-Heidelberg-New York, Springer-Verlag 1980. XVIII, 586 S., 219 Abb., DM 55,—. US $ 32.50. ISBN 3-540-09947-6

Albrecht, J. / Collatz, L. (Hrsg.), Numerische Behandlung von Integralgleichungen. Tagung in Oberwolfach, 18.—24. November 1979. ISNM 53. Basel-Boston-Stuttgart, Birkhäuser Verlag 1980. 275 S., sFr. 50.—. ISBN 3-7643-1105-3

Mittelmann, H. D. / Weber, H. (Eds.), Bifurcation Problems and Their Numerical Solution. ISNM 54. Workshop, Dortmund, January 15—17, 1980. Basel-Boston-Stuttgart, Birkhäuser Verlag 1980. 243 S., sFr. 39,—. ISBN 3-7643-1204-1

Lion, G. / Vergne, M., The Weil Representation, Maslov Index and Theta Series. Progress in Mathematics 6. Boston-Basel-Stuttgart, Birkhäuser Verlag 1980. 337 S., sFr. 30.—. ISBN 3-7643-3007-4

Helgason, S., The Radon Transform. Progress in Mathematics 5. Boston-Basel-Stuttgart 1980. VII, 192 S., sFr. 15.—. ISBN 3-7643-3006-6

Sydow, A. (Ed.), Systems Analysis and Simulation 1980. Proceedings of the Internat. Symposium, Berlin (GDR), September 1—5, 1980. Berlin, Akademie-Verlag 1980. 496 S., M 62,—. BN 762-811-0 (2182/5)

Michlin, S. G. / Prößdorf, S., Singuläre Integraloperatoren. Berlin, Akademie-Verlag 1980. XII, 514 S., M 96,—. BN 7626537 (6524)

Heitz, S., Mechanik fester Körper. Mit Anwendungen in Geodäsie, Geophysik und Astronomie. Band 1: Grundlagen, Dynamik starrer Körper. Bonn, Dümmler 1980. X, 339 S., 38 Abb., DM 48,—. ISBN 3-427-78951-9

FSC
www.fsc.org
MIX
Papier | Fördert gute Waldnutzung
FSC® C083411

Zeitfracht Medien GmbH
Ferdinand-Jühlke-Straße 7
99095 Erfurt, Deutschland
produktsicherheit@kolibri360.de